PEDAL POWER

PEDAL POWER

In Work, Leisure, and Transportation

Edited by
James C. McCullagh

With Contributions From: David Gordon Wilson
Stuart S. Wilson
John McGeorge
Mark Blossom
Diana Branch

 **RODALE PRESS**
Emmaus, PA

Printed on recycled paper

Library of Congress Cataloging in Publication Data

Main entry under title:
Pedal power in work, leisure, and transportation.

Bibliography: p.
Includes index.
1. Pedal-powered mechanisms.   I. McCullagh, James C.   II. Wilson, David Gordon, 1928-
TJ1049.P4   621.4   77-6422
ISBN 0-87857-178-7 paperback

6 8 10 9 7   paperback

Contents

CHAPTER FOUR:

CHAPTER FIVE:

CHAPTER SIX:

Acknowledgments

In many respects this book is an outgrowth of the good work done by Dick Ott, Lamar Laubach, and others who developed the Energy Cycle described in Chapter Three. I am deeply grateful to them for their technical assistance and advice.

I am also grateful to the hundreds of readers of *Organic Gardening and Farming* who provided valuable suggestions concerning possible applications of pedal power.

My special thanks to David Gordon Wilson, Stuart S. Wilson, John McGeorge, Mark Blossom, and Diana Branch who have considered the use of human power from new and interesting perspectives.

And my appreciation to Diane Gubich for her good research.

James C. McCullagh

Introduction

In this age of lasers and deep space probes, much of the muscle in the industrialized world sags like a rag doll.

In this rich technological age much of the population, particularly in developing countries, has been displaced from the workplace by "inappropriate" technologies. The leap is from bullock cart to jet plane.

Thus the paradox. Part of the world dreams of the likelihood of a "workless" state ripe with leisure time; the other part is trying to catch up. And the "catching up" is sometimes written in bold letters. On any given day, London, Lagos, and Tokyo can experience traffic jams of similar proportions.

The industrialized nations, especially America, have given birth to certain assumptions which are rapidly gaining currency around the globe: cars are "better" than bicycles; processed foods are "better" than natural ones; living in a city is "better" than living in a rural area.

Ironically, because developing countries must live by the rules of a capital-intensive economic order, they are often obliged to accept the above assumptions.

The retreat, then, is from the town, from the bicycle, from the land. The result: a death of simplicity, both in life-style and machine. Mother's milk is no longer in fashion.

Interestingly, at the time when the "appropriateness" of technology is being questioned daily, the bicycle, which is perhaps the most "appropriate" and efficient machine ever invented, is making a rocky comeback in many countries. A compelling example of this renaissance is in Dodema, the new capital of Tanzania, now under construction. The master plan for Dodema calls for a complex network of roads that will encourage maximum use of the bicycle. Furthermore, the plan decrees that the ratio of bicycles to cars will be 70:30, thus assuring this machine a major role in an enlightened transportation system.

In addition, the bicycle has come under close scrutiny by those who believe it offers great potential for performing stationary work and for goods transportation.

The literature is filled with examples of pedal- and treadle-driven machines designed to perform all types of useful work. However, with the decline of the bicycle came the decline in pedal-power devices, with the notable exceptions of apparatus used in China and other Asian countries.

Partly because of a general critique of technology, there has been in the last decade a renewed and vigorous interest in the potential of pedal power for both developing and developed countries. Professor Stuart S. Wilson of Oxford University, Professor David Gordon Wilson of the Massachusetts Institute of Technology, the Intermediate Technology Development in the United Kingdom, the

Rodale Press Research and Development Department, and many inventors, scientists, and tinkerers have brought new and valuable insights to the science of pedal power; and from these minds has come this book.

Pedal Power is a philosophical work in that it explores the full human potential inherent in the use of the bicycle for work. On the other hand, it is a very practical book, as it suggests scores of tasks which can be easily and effectively accomplished with pedal devices.

In Chapter One David Gordon Wilson charts the use of human muscle in history and explores the singular display of pedal equipment invented at the end of the nineteenth century. In Chapter Two Stuart Wilson discusses a whole range of pedal-power apparatus which could be particularly useful to developing countries.

In Chapter Three Diana Branch offers a detailed account of the Rodale Energy Cycle. The author not only gives a description of this "complete" machine, but also provides a full set of building instructions for those who desire to construct it. Furthermore, the author offers a host of suggestions about how the Energy Cycle can be effectively used around the house, garden, farm, and homestead.

The possible uses for pedal power are significantly extended by John McGeorge and David Gordon Wilson in Chapters Four and Six, respectively. And in Chapter Five Mark Blossom considers "Treadle Power in the Workshop."

Overall, the book is ripe with plans, models, prototypes, and possibilities. It is hoped the reader will be moved to develop pedal equipment of his own.

Above all, we hope this book will bring about a reconsideration of the bicycle and pedal-driven machines. We feel these are subjects worthy of additional study.

Perhaps an interface between East and West is the bicycle, the machine which makes us all brothers and sisters.

SHANTUNG PLOW

**Two sticks
are matches for striking
soil at man's fertility,
sweat at his crotch
his sun.**

James C. McCullagh

CHAPTER ONE
Human Muscle Power in History

By David Gordon Wilson

"The sweat of the brow is daily expended by millions, and daily millions of sighs are wrung from the tormented frame of the bent and weary in the pursuit of providing food." Rudolf P. Hommel wrote this after living for eight years in China in the 1920s studying Chinese tools and crafts. His aim was to "give a fairly complete picture of Chinese life, as lived by millions of people today, a life in which there has been no considerable change for thousands of years."

That picture is, I believe, one which we all have of our forebears in any culture except, perhaps, those few tropical paradises where we are taught to believe that the inhabitants just sat under the banana trees and coconut palms eating their fruits whenever they wished. My remembrances of growing up in England in the thirties and forties are certainly closer to the Chinese model than to that of the Pacific islands. During World War II we all had large vegetable gardens carved out of tennis courts and the like, and I look back without longing at the back-breaking weeding, watering, and the "double-digging" (double-depth trenching the plots, with manure in the lower part of the trenches). The only mechanization we had was my homemade bicycle trailer which carried the

five-gallon cans of water and the manure. The fork, the spade, and the hoe were the principal tools; and they used, or misused, our bodies painfully. We could utilize only a small proportion of the energy output of which we were capable because of the twisting contortions which these implements demanded of us. How different from the relative comfort of a bicycle, with a choice of gear ratios to suit the load and the terrain.

I have worked on farms in England, Scotland, and Germany and have lived among farmers in Nigeria. In all these places the tractor was beginning to take over those tasks which could be most easily mechanized. But this meant that the manual labor which was left to be done was generally the least susceptible to relief given by the application of mechanical aids. We shoveled endless quantities of manure; we hoed the weeds from almost invisible crops; and we picked up potatoes from the mixture of earth and stones thrown up by a speeding tractor with a rotary digger. We did not feel that we were much better off than our more ancient ancestors.

What is remarkable about the historical use of muscle power is not only how crude it generally was, but that when improved methods were tried, they were generally not copied and extended. There were three ways in which the application of human muscle power could fall short of the optimum. First, the wrong muscles could be involved. We find time and again that people were called upon to produce maximum power output, for instance in pumping or lifting water from a well or ditch, using only their arm and back muscles. It seems obvious to us nowadays that to give maximum output with minimum strain we must use our leg muscles, not incorrectly called our second heart.

Second, the speed of the muscle motion was usually far too low. People were required to heave and shove with all their might, gaining an occasional inch or two. A modern parallel would be to force bicyclists to pedal up the steepest hills in the highest gears, or to require oarsmen to row boats with very long oars having very short inboard handles.

Third, the type of motion itself, even if carried out at the best speed using the leg muscles, could be nonoptimum in a rather abstruse way. Here is the best example I know of: Dr. J. Harrison of Australia took four young, strong athletic men and a specially built "ergometer"—a device like an exercise bicycle in which the power output could be precisely measured. He wanted to settle the controversy as to whether oarsmen produced more or less power than bicyclists, and he reproduced the leg and arm motions required for rowing racing boats (or "shells") and pedaling racing bicycles. He found (somewhat to his surprise, no doubt) that there was negligible difference between the power output produced in these two very different actions by the same athletes after they had practiced long enough to become accustomed to each.

Then he tried some old, and some possibly new, variations. He fitted elliptical chainwheels in place of the normal circular types to the cranks of the bicycle-motion devices. These chainwheels were made in Europe in the thirties to reduce the apparently useless time spent by the feet at the top and bottom of the pedaling stroke in bicycling, and correspondingly to increase the more useful time when the legs are going down in the "power" stroke. He found that some of his subjects, but not all, could produce a little more power with the elliptic chainwheel than they could before. Then he changed the ergometer so that, instead of the rowing motion usually found in racing boats where the feet are fixed and the seat slides back and forth, the seat was fixed and the feet did the sliding. This time all his subjects produced a perceptible increase in power output. The reason was apparently that they did not have to accelerate so high a proportion of their body mass at each stroke.

In normal rowing, after the oarsman has driven the oars through the water by straightening out his legs and body, he must then use muscles to eliminate the kinetic energy produced with such effort in the body. Harrison investigated the effects of using mechanisms which automatically conserved this kinetic energy. He used various types of slider-crank motions, like those of a piston in an automobile cylinder. He called these "forced," as opposed to the normal "free," rowing motions; and he found that all his subjects produced a substantial increase over their previous best power outputs in rowing or bicycling. What is more, this improvement held for as long as the tests went on. One subject, apparently Harrison himself, could produce no less than 2 horsepower (1.5 kilowatts of mechanical output) for a few seconds, and a more-or-less continuous output after five minutes of a half horsepower, still 12 percent or so above his best output by other motions.

This careful, scientific work enables us to look with a better perspective at the use of human muscle power in the past. Until Harrison did his work, no one could agree as to which muscle action was best to use for racing or for steady, all-day work. Even now, six years after wide publication of his results, no one to my knowledge has grasped the significance sufficiently to apply this new information to ease the lot of anyone who has to use muscles in his daily work or to increase the speed of people who race. And, incidentally, other research by Frank Whitt in England has shown that power output measured by ergometers may be substantially lower than that produced by the same persons using the same muscle actions when bicycling or rowing because the absence of the self-produced cooling wind results in dangerously overheating the body. As pointed out earlier, few of the motions used historically to harness human muscle power incorporated any intrinsic cooling action. They were mostly of the slow, heaving variety, so that our unfortunate forebears had to cope with heat stress on top of the use of usually inappropriate muscles moving against resistances which were too large at speeds which were too low. If in the future we run out of the earth's stored energy and have to resort to that of our bodies, we should be able to look forward to considerably greater comfort while we are working—if the results of modern research are applied.

The Manpower Plow of Shantung

This was, and possibly still is, a plow operated by two men, one pushing and one pulling. Rudolf Hommel found it still being used in China in the twenties. "Shantung is very much overpopulated, and poverty is therefore much in evidence. . . . [I]t is therefore not surprising to find today a primitive plow, which for lack of draft animals has to be served by man to pull it. There is a baseboard with a cast-iron share at one end. Two uprights are firmly mortised into the baseboard, the rear one of which, farthest from the share and bent backward, resem-

Figure 1-1 The Shantung plow

bles the handle of one of the ancient one-handled European plows, but is not so used. Instead of grasping the upper end of this upright in his hands, as in the old western plow, the plowman, leaning forward and downward, presses his shoulder against it, while his two hands grasp the two projecting ends of a cross peg-handle driven through the lower part of a curved upright. Thus in a very ingenious manner, he not only guides but pushes the plow."

For this arduous task, both plowmen used their leg muscles, which are the most appropriate muscles for the duty. The motions are too slow to be efficient (in engineering we call this a poor "impedance match"), and most of the other muscles and body frame are strained painfully to apply the force produced by the leg muscles. One hopes that it was used only in soft ground. In the rocky soil of New England its use would be exquisite torture.

I have started with this plow because we have so good a description of it, complete with a knowledge of how it was used. In most historical cases, we have just old illustrations which were made for purposes other than for showing the details of the mechanisms or the precise way in which they were used. We can usually guess intelligently enough. But before we leave the manpower plow, consider how you would perform the same task today. I know of no purchasable alternative to the fork and spade—for either of which my back has no great affection. Certainly the Rodale winch described in Chapter

Three is a solid advance. We will be discussing various other alternatives in the chapter on futuristic uses of manpower.

In the examples which follow, I am not attempting historical completeness: I have chosen them as interesting illustrations of how muscle power has been used in the past for a variety of tasks. I am grouping them by the muscles and motions employed.

Handcranking

This is perhaps the most obvious means of obtaining rotary motion, and man has been using it for centuries. The earliest known handcranked device was a bucket-chain bilge pump found on two huge barges used by the Roman emperors and uncovered when Lake Nemi was drained in 1932.

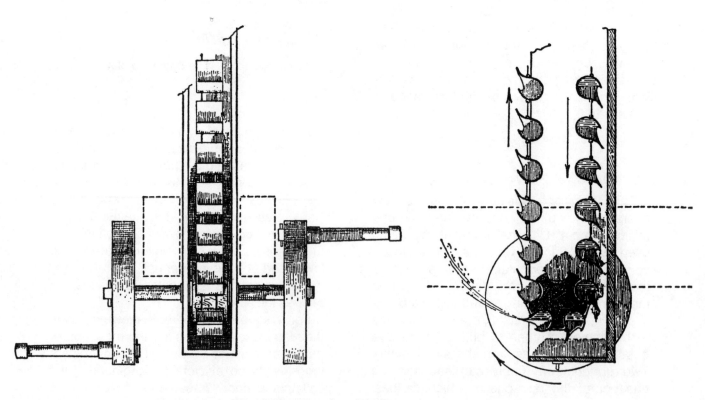

Figure 1-2 The bucket-chain bilge pump (Reproduced by permission of Doubleday & Company, Inc.)

Figure 1-3 Bucket-chain water lifter (Courtesy of Friedrich Klemm)

Figure 1-4 Chinese endless-chain water lifter (Courtesy of Martha Hommel)

Agricola, writing in 1556, showed a complicated handcranked transmission for driving a similar bucket-chain water lifter. He also showed a bucket-chain being assembled. An endless-chain water lifter was also used in China in much later times. It was different in two respects. Instead of buckets, the water was trapped by boards sliding in a trough. One would think, however, that this would be less efficient because of friction and leakage. In addition, levers were attached to the cranks, with all the lost motion and top-dead-center problems they entailed. Presumably the levers were used to give a more comfortable working position for the ground-mounted trough.

Leonardo da Vinci shows concern for the comfort of the user in his drawing of a textile winder with a handle at a convenient height and with a winder-drum of a diameter giving what will presumably be a near-optimum rate of action. Leonardo uses gearing for the same reason—obtaining a good "impedance match"—in his design of a file-cutting machine in which the crank is used to raise a weight at a speed to suit the operator, and the weight subsequently delivers energy at an optimum rate to the drop-hammer cutter.

An earlier crank-driven screw-cutting lathe was obviously not designed by Leonardo.

One can imagine the difficulty of simultaneously turning a high-resistance load with a small crank in one hand while trying to control the cutting process with the left hand.

Two much more modern examples of handcranking are taken at random from the

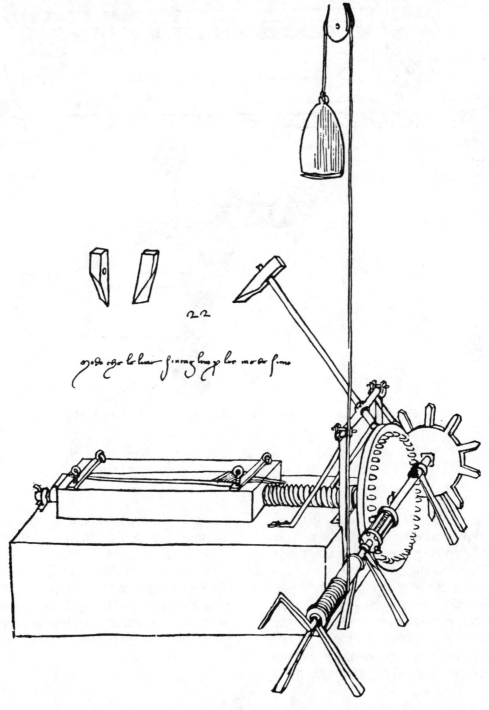

Figure 1-5 Leonardo's file-cutting machine (Courtesy of Friedrich Klemm)

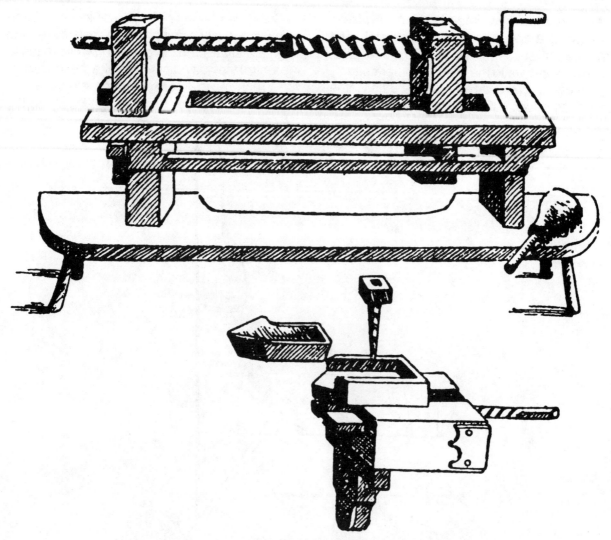

Figure 1-6 Screw-cutting lathe (Courtesy of Friedrich Klemm)

Science Record of 1872: the air pump for an undersea diver and what looks like a multiple stirrer for a nitro-glycerine-manufacturing process. These seemed to be low-torque applications of muscle power. A high-torque application which scarcely needs illustration was the old hand-wringer, which I used to try to turn for my mother. This was rather similar to the fifteenth century screw-cutting lathe in that while the right hand turned a heavy and fluctuating load, the left hand had to perform a difficult and hazardous control function.

A variation of the handcrank was used in China in the form of a "T-bar" attached to the crank. The use of this simple connecting rod enabled the use of both hands and/or one's chest or belly to contribute to overcoming the resistance.

READY TO GO DOWN.

Figure 1-7 Air pump for undersea diver

MAKING NITRO-GLYCERINE.

Figure 1-8 Nitro-glycerine factory

Levers Actuated by
Arm and Back Muscles

Until the arrival of the sliding-seat scull, oars were moved predominantly through the action of the arms and back. Battles among warships were won by the boat which could pack in the most oarsmen. Ameinokles of Corinth in about 700 B.C. built boats to accommodate three rows of oarsmen in a staggered arrangement on each side; with almost 200 oarsmen, it could travel at seven knots and became the standard battleship of the Mediterranean.

At the other end of the warlike scale were the pipe organs designed by Ktesibios in Alexandria in the third century before Christ. The air pump was a rocking lever which could be operated with two hands. There was little difference in the external appearance, at least, from the hand-pumped organ used in our church in England when I was a boy. (My father fitted it with one of the first electric blowers used for the purpose, at least in our area of the country.)

Some tricycles were designed for lever propulsion by the hands, the arms, and possibly the back. Sharp showed a drawing

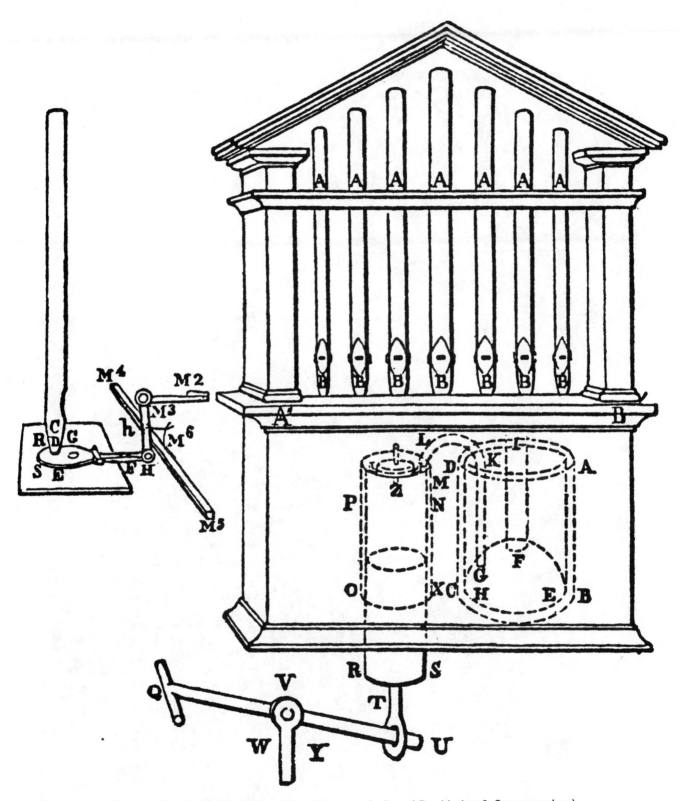

Figure 1-9 Pipe organs (Reproduced by permission of Doubleday & Company, Inc.)

Figure 1-10 Singer "Velociman"

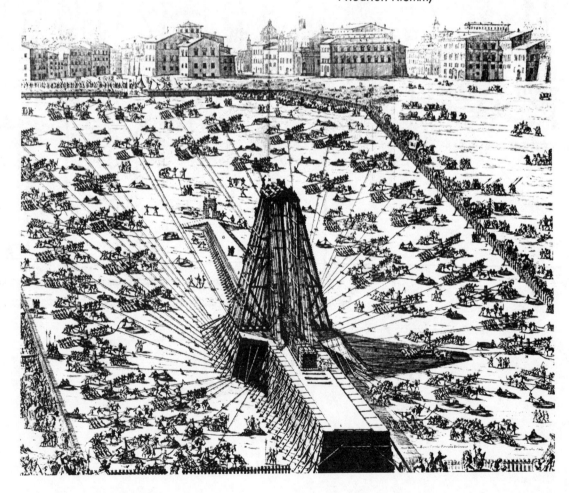

Figure 1-11 Erection of a massive obelisk at the Vatican (Courtesy of Friedrich Klemm)

of a Singer "Velociman." The arms pulled two levers which pivoted on swinging arms and operated the cranks of a transverse shaft. This drove the wheels through a chain drive. The driver steered by pivoting the back rest. With this type of drive, the legs could atrophy, as has been predicted if modern automobiles become any more automatic. This vehicle flourished in the 1880s and 1890s. Much earlier than that, in 1821, Louis Gompartz made a velocipede in which propulsion and steering were supposed to take place through swinging a lever over the front wheel. The lever carried an arc with a rack gear that engaged a circular gear, presumably on a free-wheel, on the front hub.

Capstans

The windlass or capstan represented an enormous improvement over most of the foregoing devices when maximum work output was to be given by human muscle power. It involved the large muscles of the legs. The motions were those of walking, which must be at least reasonably efficient; and the motion speed could be varied simply by using a smaller or a larger winding-spool diameter. It seems likely, however, that in fact slow, high-force pushing was used more often than the fairly rapid light-force walking which modern research would show to be much more efficient.

Capstans have probably been in use almost as long as there have been ropes. A literally monumental use of capstans was in the erection of a 360-ton obelisk at the Vatican in Rome by Pope Sixtus V in 1586. Forty capstans were used, as well as 140 horses and 800 men, in a military-type operation.

Treadmills

Of the devices so far discussed, treadmills are the nearest approach to true "pedal power." They seem to be at least as ancient. Varieties of treadmills were in use in Mesopotamia 1200 years before Christ. They continued in use in Europe at least until 1888, when the last treadmill crane on the lower Rhine ceased operation, but they were still being used in China at the time of Rudolf Hommel's stay there in the 1920s.

Treadmills had the same advantages as capstans. The motion was walking, and the gear ratio could be easily adjusted to be near optimum (but probably seldom was). Some treadmills shown in Agricola's 1556 book on mechanics and mining looked exactly the same as capstans except that the radial handles were fixed and the circular walkway rotated. No doubt this improved comfort and performance in long-duration work because the power plant was less likely to become giddy.

A variation of the capstanlike treadmill was one with the rotating footwheel inclined. Rather than continually pushing on a bar to force their feet back, the men on an inclined treadmill moved as if they were climbing an endless ramp or flight of stairs. Their weight was usually enough to carry the wheel around, and the horizontal bar was less for pushing against than for steadying the operator(s). One disadvantage of the inclined treadmill was that, whereas in a horizontal mill several men could all push at once as on a capstan, only one person could be located in the optimum position of an inclined treadmill.

Most treadmills were of the squirrel-cage variety, and, in fact, various animals from dogs to horses were frequently used. But when tasks requiring close control, such as lifting weights in a crane for building construction, for instance, were called for, men were more usually employed. Thus men worked treadmills in winches and cranes, and animals were used in devices powering irrigation pumps or forge blowers.

Leonardo da Vinci, as usual, has the last word on ingenuity. He designed a treadmill to

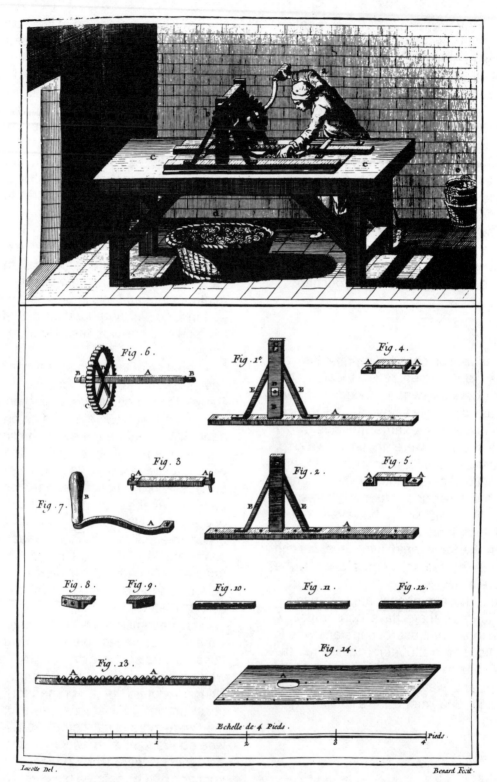

Monnoyage, Machine pour la Marque sur tranche

Figure 1-12 Handcranked machine used in marking metal coins
(Compliments of Lehigh University, Honeyman Collection)

bend and cock four crossbows mounted radially on the inside of the wheel. A single archer loaded and fired each in turn: it was a type of Gatling gun or revolver. The wheel had unidirectional, comfortable-appearing steps on the outside of the wheel for the several people to use when supplying the motive power. This must have been a superior position to being cooped up inside a wheel in an often-cramped position. Leonardo took pains, too, to protect the workers as well as the archer with armor. Most of his designs were never made; we don't know if this repeating crossbow was, but it is doubtful.

We can see from the accompanying examples from *The Encyclopedie des Planches* (1751) by Denis Diderot and Jean Lerond D'Alembert that the European workshop of this period used handcranking devices to a considerable extent. While in some cases the method appears to fit the task, in others, such as in the scenes from an optician's studio and from a knife-cutting shop, handcranking appears to be woefully inadequate. On the other hand, the eighteenth century workshop had some very sensitive machines, such as the one for engraving fine stone. Note the application of the treadle and the flywheel.

Legs on Treadles

The application of leg muscles to treadles can be roughly divided into two categories: that where low power was required and the hands were required to perform an accurate task, such as in treadle sewing machines, and those in which maximum power output was desired, as in certain types of cycles.

Treadles often provided energy in the low-power category in a reciprocating manner. We have, for instance, illustrations of a bowstring boring machine used for drilling pearls for necklaces and a foot-operated lathe with bowstring and overhead cantilever spring, both dating from the fourteenth century. The Chinese used treadles to obtain continuous motion for cotton ginning and spinning, but the treadles were almost cranks. One end of the treadles was mounted in a universal bearing at near floor level, while the other was fitted in another universal bearing eccentrically on the drive wheel. Apparently both feet would be used on one treadle, because the wheel would generally be inclined to give a favorable angle to the treadle on one side and would make the use of a treadle on the other side very difficult.

The use of treadles for maximum power output has been principally in their application to cycles. They were generally connected to cranks on the driving wheel. The first pedaled bicycle, made by Alexander Macmillan in the period 1839–1842, was of this type. Tricycles and four-wheelers often used this system. But the American Star, which was introduced in the 1880s as a safer version of the "ordinary" or "penny-farthing," used a strap going from the foot-levers to a spool on the wheel. The spool was mounted on a one-way clutch or free-wheel, with a spring tending to wind up the strap. The rider could push down on the levers together or alternately to propel the bicycle forward. The diameter of the spool controlled the gear ratio, so that there was, in fact, no need for the manufacturers to use a dangerous high wheel at all. The American Star was a promising development which was, however, eclipsed by the small-wheeled chain-driven pedal-and-crank "safety" bicycle.

Leg Muscles Used in Cranking

Just as the "high-wheelers" were eclipsed by safety bicycles, so the lever propulsion systems which some of them used disappeared, and during the 1890s the pedal-and-crank drive became almost universal. The essentials of the safety bicycle in almost all its aspects had been developed by the turn

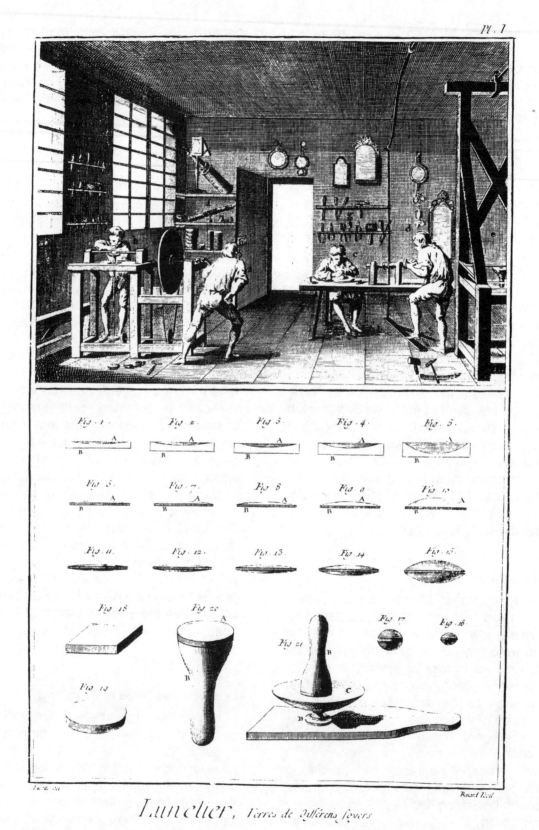

Figure 1-13 Optician's studio (Compliments of Lehigh University, Honeyman Collection)

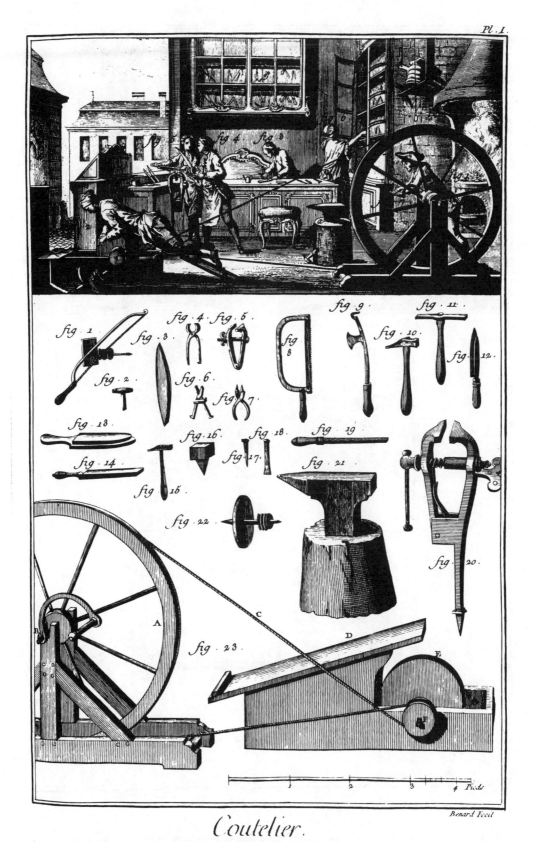

Coutelier.

Figure 1-14 A machine for cutting and sharpening knives
(Compliments of Lehigh University, Honeyman Collection)

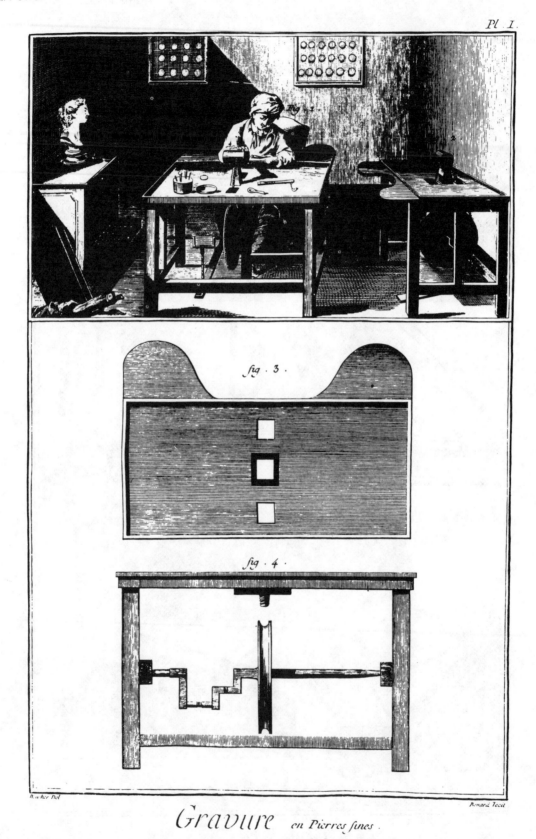

Figure 1-15 A machine for engraving fine stone (Compliments of Lehigh University, Honeyman Collection)

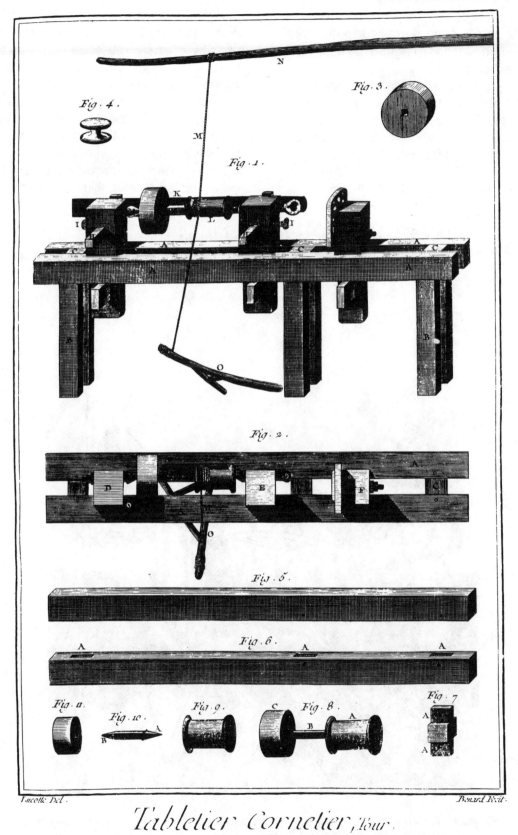

Tabletier Cornetier, Tour.

Figure 1-16 Treadle-powered lathe for making spools of thread (Compliments of Lehigh University, Honeyman Collection)

of the century—even derrailleur gears had arrived—and it has reigned almost unchallenged since then.

One can believe, with the benefit of hindsight, that so obvious a system as pedals and cranks must have been used for muscle-power applications before the advent of the bicycle, but I have been able to find no record of them. It is possible that the design of a cantilevered pedal with low-friction bearings to take the large forces which can be applied by a heavy, muscular man was too difficult for the low-strength materials which were available earlier.

Once pedals and cranks had been developed for bicycles, however, they began to appear in many other applications. Racing boats were built and were easily able to beat those manned by trained oarsmen. Pedals were applied to tools like lathes, saws, and

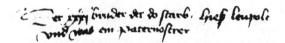

Figure 1-17 Bowstring-operated boring machine for preparing pearls for necklaces (Courtesy of Friedrich Klemm)

pumps. Many attempts have even been made to fly a man-powered airplane. The propulsion system in every known case has been by standard bicycle-drive components.

Whether the standard bicycle arrangement is optimum is often a matter of heated debate. The proponents of the standard system reject all challengers. In 1933 a so-called Velocar was introduced in France. It was a bicycle in which the rider sat, or lay, behind rather than over the cranks. It "broke all cycling track records" according to a news story "but was later banned by the International Cycling Union on the grounds that it was not a bicycle." We are lucky to have people whose minds are more open nowadays. Dr. Chester Kyle, running what has become an annual world speed-record event at Long Beach in California, allows any type of human-powered vehicle to enter. The variety

Figure 1-18 Foot-powered lathe (Courtesy of Friedrich Klemm)

Figure 1-19 Chinese treadle "crank" (Courtesy of Martha Hommel)

Figure 1-20 The American Star

Figure 1-22 Velocipede powering a sewing machine (U.S. Bureau of Public Roads, Photo No. 30-N-41-356)

BICYCLE EXERCISES AT LEIPZIG.

Figure 1-21 Bicycle exercises at Leipzig (U.S. Bureau of Public Roads, Photo No. 30-N-33-44)

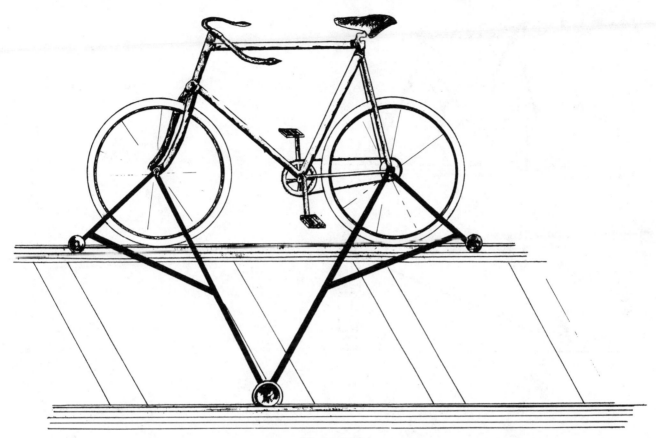

Figure 1-23

Figures 1-23 and 1-24   Renditions of two of the numerous patents for railroad bicycle attachments

of the initial contestants' machines has been extraordinary. Out of these contests there may emerge methods of using the leg muscles with greater effectiveness, not only for superathletes competing for the glory of being fastest, but also for those who labor at more mundane intensities.

Pedal Power in the Workshop

With the invention of the bicycle came a veritable avalanche of pedal and treadle ma-chines. The bicycle influenced all aspects of life: work, sports, leisure, and transportation.

Errand boys, policemen, and post office workers discovered that the bicycle made them mobile or more efficient. Women, so long tied to the home and garden, found some "liberation" in the two- and three-wheeler. And the sports world, which wor-shipped the speed of the trotter and race horse, turned to the bicycle racer, who promised undreamed-of speeds.

Resourceful Americans and Europeans tried to adapt the principle of the bicycle to all

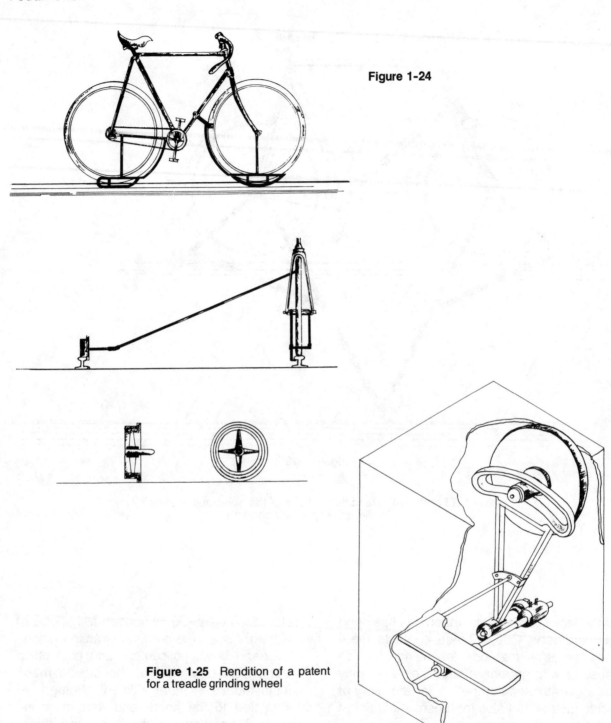

Figure 1-24

Figure 1-25 Rendition of a patent for a treadle grinding wheel

Figure 1-26 Primitive cotton factory in Alabama—ca. 1890 (Culver Pictures, Inc.)

parts of the home and workplace. Some applications, depicted in the etching of an "exercise" velocipede, were fanciful. But others were not. Consider the drawing of a velocipede powering a sewing machine. The artist, who was perhaps parodying the high-riding racer, reveals quite accurately the "work" potential of the bike. And consider the patents for railroad bicycle attachments, a concept which is being revived today (see chapter six). The patent office was alive with designs for countless applications for pedal power.

All in all, the bicycle seemed to present the possibility of humanizing the workplace, of relieving men and women of some of the drudgery associated with arduous tasks.

It is unlikely that we will ever know the full story of "pedal power" at the turn of the century. We do know, however, that national mail-order companies, including H. L. Shepherd & Co., W. F. Barnes & Co., and the Seneca Falls Machine Company, lost little time in applying features of the bicycle to uses in the home and workplace. The accompanying catalog reproductions of

NO. 1 AMATEUR SAW.

This machine will cut pine of any thickness up to 1½ inches, and harder woods of proportionate thicknesses. It admits a swing of 18 inches around the blade, and accomplishes every branch of sawing within the range of general amateur work.

The table does not tilt, as in No. 6 Amateur Saw, but sawing for inlay work can be done by placing a beveled strip under the stuff being sawed.

The price of the machine complete is $10.00.

The price of the machine without Boring Attachment is $8.00.

It weighs 40 pounds.

Boxed ready for shipment it weighs 60 pounds.

NO. 6 AMATEUR SAW.

We offer this machine, believing it to embody all that the most critical can desire. It will cut pine of any thickness up to one inch, and harder woods of proportionate thickness. It admits a swing of 16 inches around the blade, and will saw and drill ivory, bone, metal, shells, etc.

The blade has hardened steel clamps on slides which move in permanent guideways, above and below the table, giving a positive and accurate motion.

The table has a beveled adjustment, whereby it can be set for inlaying, mosaic and other work.

The price of the machine complete is $12. The price without Boring Attachment is $10.

It weighs 40 pounds. Boxed ready for shipment it weighs 60 pounds.

We include one dozen blades with each machine.

Figure 1-27 "Amateur" saw, 1892 (Eleutherian Mills Historical Library)

SCROLL SAW NO. 7 IMPROVED.

(Formerly Called Large Size Scroll Saw.)

Price, $15.00.

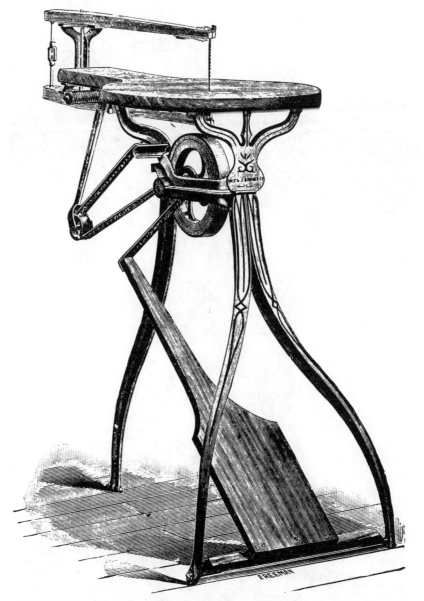

Figure 1-28   Scroll saw, 1892 (Eleutherian Mills Historical Library)

VELOCIPEDE SCROLL SAW NO. 2.—Improved.

Price, Complete, $23; Without Boring Attachment, $20.

Figure 1-29 Velocipede scroll saw (Eleutherian Mills Historical Library)

Human Muscle Power in History

FOOT POWER FORMER.—Improved.

Price $20.00.   Knives Extra.

Figure 1-30 Grinding wheel (Eleutherian Mills Historical Library)

Lathe No. 4.

7-Inch Swing.

Price, $40.00.

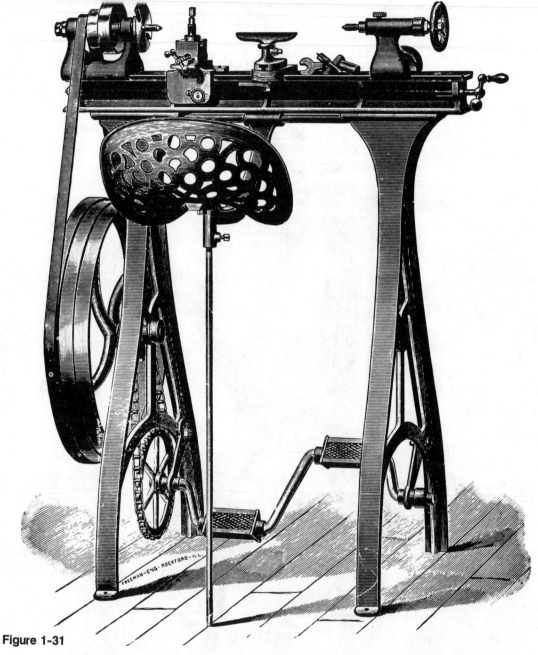

Figure 1-31

When ordering lathes, be particular to state clearly whether wanted with foot power or countershaft; if with foot power, state whether velocipede or treadle.

Figures 1-31 through 1-34   Pedal-powered lathes and descriptions
(Eleutherian Mills Historical Library)

Lathe No. 4.

7-Inch Swing.

THIS lathe is designed for turning both wood and iron, and for boring drilling, polishing, etc. It is a desirable tool for small work, and has many important advantages in the construction and arrangement of its parts. It swings 7 inches and takes 20 inches between centers. It has our patent velocipede foot power, which is the best power ever applied to a foot-driven lathe. The speed can be varied from 1,000 to 2,000 revolutions per minute, and the motion can be started, stopped or reversed instantly, at the will of the operator. Greater power can be applied on the work than with any old-style foot power, and with greater ease. The lathe is made entirely of iron and steel. The bed is solid, and has V-shaped projections, over which the head and tail stocks and hand and slide rests are fitted. The lead screw for the carriage is operated by hand; by it the carriage can be traveled 20 inches, the entire distance between centers. The carriage can be engaged or disengaged instantly from the lead screw. The cross feed way on which the tool post moves can be set at any desired angle for taper turning and boring. The tail stock can be moved and set at any point desired by the simple turning of the hand-wheel; or it can be taken off entirely, thus leaving the bed free for face-plate or chuck work. The head stock spindle is hollow, size of hole $\frac{9}{32}$ inch. The head stock spindle has taper bearings, and is capable of very nice adjustment. The tail stock center is self-discharging.

The price of the lathe is $40.00; this includes face-plate, two pointed centers and one spur center, hand rest, wrenches and necessary belting, as shown in cut.

The lathe weighs 210 pounds.

Boxed, ready for shipment, 265 pounds.

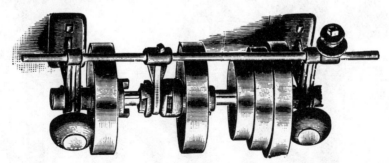

The above cut represents a countershaft for No. 4 and 4½ lathes.

The pulleys on countershaft are 7x1½ inches, and should be speeded 250 revolutions.

Price of countershaft, $15.00.

We can furnish lathe with countershaft in place of foot power at same price as with foot power.

Figure 1-32

Screw Cutting Lathe No. 4½.

9-Inch Swing.

Price, $65.00.

When ordering lathes, be particular to state clearly whether wanted with foot power or countershaft; if with foot power, state whether velocipede or treadle.

Figure 1-33

Screw Cutting Engine Lathe No. 6.

13-Inch Swing.

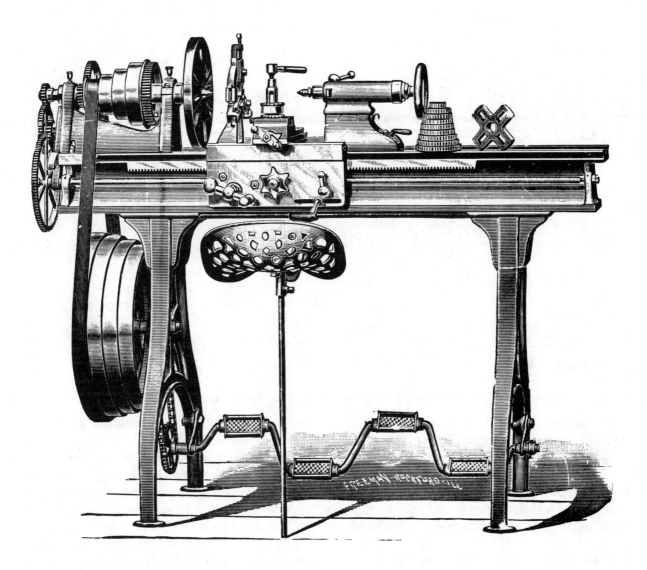

When ordering lathes, be particular to state clearly whether wanted with foot power or countershaft; if with foot power, state whether velocipede or treadle.

Figure 1-34

4

able gunsmith's lathe it has ever been our good fortune to see. The machine is already set up ready for operation and is a marvel of ingenuity. It is self-acting, back geared, and is so complete in its working as to be able to cut screws from the smallest required by a gunsmith, to the largest. This is the only machine of the kind in Canada, and Mr. Soper deserves great credit for his enterprise.

———

IN REFERENCE TO THE $50, SCREW CUTTING LATHE.

PORT HURON, MICH., Dec. 26, 1876.

MR. H. L. SHEPARD,

Dear Sir: I have delayed writing to you before as I wished to try the Lathe before saying anything about it. It answers all my expectations and is the best Lathe I ever saw for the money. Several mechanics (hearing I had the Lathe) have called to see it and all speak favorably of it. You may expect 2 more orders soon.

Respectfully yours,
C. W. DANGER.
U. S. Survey, Port Huron, Mich.

———

AUSTIN, ILL., Dec. 26, 1876.

MR. H. L. SHEPARD,

I am very badly disappointed in my Lathe. Taking the Price into consideration, I expected to get a light build shaky machine, but instead I have a really good small size Engine Lathe. The Self Feed for Metal Turning is a great feature, and works splendidly. I pretty thoroughly tested its value in turning out a Spur Center from a piece of inch steel, which I did without difficulty. I consider it a better Tool than I ever saw before for twice the money, and am entirely satisfied with it. You may use my name in any way you choose, and refer to me and I will be happy to answer by mail any inquiries that may be made, and will also exhibit the Lathe to any one calling on me. Hoping I may influence others to patronize you, and wishing for your entire success in your enterprise, I remain yours very truly,

AL. J. FARRELLY, P. M.
Austin, Cook Co., Ill.

5

LATE TESTIMONIALS WHICH SPEAK FOR THEMSELVES.

———

CAN SCARCELY FEEL FOOT MOTION WHEN CHASING.

MORRIS, ILL., Dec., 24, 1876.

MR. H. L. SHEPARD,

Dear Sir: The Screw Cutting ($50.) Lathe with Independent Rod-Feed you shipped me arrived in due time, after using it I find it will do all you claim for it and more. I have been looking and making inquiries for the past three years for a Screw Cutting Lathe, but could not find one to suit me until I saw yours. The Lathe is well made and I see you have not spared a little extra expense in the use of Gun Metal and Steele where it was of use for wear and ease of movement. It is well Braced and Strong without being clumsy, making a handsome as well as a useful Tool. The Self Feed works very easy and smooth, I think you have made a decided success of your propelling power. It runs very easy with the Screw Cutter thrown in. You would scarcely know you were running a Lathe by foot power, and I think any one in want of a Foot Lathe cannot do better than yours. Hoping you will be sufficiently paid for the benefit you have and will do users of your Lathe, I remain yours,

D. S. HUFF.

———

TESTIMONIAL—$25 LATHE.

LA CROSSE, WIS.

MR. H. L. SHEPARD,

Dear Sir: The Lathe arrived all right and is just what you advertise it to be.

Respectfully yours,
E. T. MILLER.

———

PORTLAND, MAINE, Dec. 20, 1877.

MESSRS. H. L. SHEPARD & CO.

Gents: The $150. Lathe which I purchased of you, gives satisfaction. In style, design, and working parts, is well worth the price. I have thoroughly tried it in turning, screw cutting and milling, and would most cheerfully endorse and recommend it to those wishing a Foot Lathe.

Yours Respectfully,
C. F. DAM.
269½ Congress Street.

Figure 1-35 Letters from users of foot-powered machines (Eleutherian Mills Historical Library)

foot-powered lathes, saws, and grinding wheels reveal a remarkable simplicity of design. Some of the machines seem very "modern" in design. The screw-cutting lathes represent a genuine maturity of design which is not far removed from that of the machines used in countless workshops today.

If the bicycle represented a revolution of sorts, so did the manufacture of foot-power tools which inaugurated, in a small way, the home workshop, where even the unskilled could perform certain machine tasks with accuracy.

Unfortunately, the internal combustion engine retired, for the most part, bicycles and pedal-power machines. The bicycle design was not improved; few explored additional pedal-power possibilities.

For close to a century in the industrialized world the application of human power to transportation and useful work was seldom contemplated. However, decades of mechanization and pollution have led us to reconsider human muscle potential. Ironically, as developing countries strain to give up their bikes and pedal-power equipment, industrialized countries are giving more and more attention to these items. Hopefully, all countries will reconsider the enlightened use of pedal power. The high costs of energy and the failure of our transportation systems make such reconsideration imperative.

36

ARCHIMEDES SCREW

**Turn the river
on its back
dig a peep hole
to China.**

James C. McCullagh

CHAPTER TWO
Pedal Power on the Land:
The Third World and Beyond

By Stuart S. Wilson

It is arguable that the most important development, technically and socially, in the nineteenth century was the bicycle. Pre-bicycle technology was heavy and inefficient—typically the steam locomotive—but post-bicycle technology became lightweight and efficient, both structurally and mechanically. Witness the lightweight tubular steel frame, wire spoked wheel, bush roller chain, ball bearings, and pneumatic tire. All these, developed specifically for the bicycle, led to a triumph of ergonomics—matching the machine to the human being—which explains why the bicycle has succeeded so widely in providing cheap and effective personal mobility for all, worldwide.

Unfortunately it also led to the development of the automobile, which has had such a disastrous effect on our way of life in the twentieth century, yet the signs today are that the automobile has passed its peak while the bicycle continues its steady upward trend—a tortoise and hare situation.

In energy terms the reason the bicycle is so efficient is that it uses the most powerful muscles in the body—the thigh muscles—in the right motion, a circular pedaling motion, at the right speed, 60–80 revolutions per minute, and then transmits the power efficiently by means of a sprocket-and-chain mechanism and ball bearings.

The torque or turning effort exerted by the feet on the crankshaft and therefore on the sprocket and chain are not constant, being appreciably smaller *but not zero* at the top and bottom positions of the pedals. This minimum torque is achieved partly by "ankling" (tilting the top foot upwards and the bottom foot downwards), partly by friction of the shoes on the pedal rubber, or by the use of toe clips. The minimum torque varies with individuals but is of the order of one third of the maximum. On a bicycle this variation has little effect because of the inertia of the rider and machine; but if we consider driving a stationary machine, for example a pump or a corn grinder, then the motion becomes jerky and it is desirable to even it out, either by incorporation of a flywheel or by other means, such as the use of an elliptical sprocket, which in effect varies the gear ratio twice during each revolution of the crank.

How much power can we expect to be able to supply by pedaling? Tests done under laboratory conditions do not relate too well to experience on the road, but tests at Oxford on a bicycle with a built-in dynamometer confirm that 1/10 horsepower or 75 watts is a reasonable figure for the sustained output of an average rider at a road speed of 12 mph. On the other hand, ¼ horsepower, or nearly 200 watts, is produced at 18 mph, which many riders can achieve at least for a limited time; and up to 1 horsepower or 750 watts is possible for a second or so.

These figures reveal a remarkable overload capacity of the human body—the ability to exert 10 times our normal output when required. The question is how to apply this muscle power to useful ends other than personal transport as with the usual bicycle. In particular, how can we develop ways and means of helping people to help themselves, in rich and in poor countries, by their own efforts, without depending on expensive oil fuel?

Transportation

Let us first consider extensions of bicycle technology to goods transport. The carrier bicycle with a large basket over the front wheel is effective for loads of up to 100 lb. or

Figure 2-1   Tricycle used for transportation of goods in Taiwan

Figure 2-2 Bicycles remain a popular transportation vehicle in Taiwan and many other countries.

so. Beyond that, however, three wheels are better than two and have been in use since the nineteenth century, but have hardly evolved. A modern attempt to rethink the design is shown in Figures 2-4 through 2-8. It was designed by the author and built at Oxford with support from Oxfam, the well-known international charity. Named the "Oxtrike," it is designed as a basic chassis with a choice of bodies to carry a variety of goods or people up to a payload of 330 lbs. (150 kg or 3 cwt). It incorporates a number of novel features in order to overcome some of the limitations experienced with existing designs of cycle rickshaw (pedicab, becak, or trisha) in current use in India, China, Indonesia, and Southeast Asia. Although a variety of designs have evolved in these different locations, there is little evidence of radical redesign in order to improve performance.

The main defects of existing designs include the use of only a single gear ratio which is often too high, so that starting on the

Figure 2-3 Foot-powered trolleys are a common form of transportation in the Philippines.

level or climbing a gradient imposes a severe strain on the driver; usually only one wheel is driven, while braking also is confined to one wheel—a dangerous defect. The use of standard bicycle parts often results in insufficient strength for the far greater loads imposed, leading to the collapse of wheels or forks unless specially strengthened. The frame itself, of tubular construction on bicycle lines, can be regarded only as a less than optimum design with regard to strength-to-weight ratio.

Despite these defects, the various types of tricycle have established themselves widely in Asia, though not in Africa, but they are fighting a losing battle in some of the bigger cities, such as Singapore and Jakarta. Some authorities wish to banish them on the grounds of lack of safety, causing congestion, and lack of a modern image. To abolish such a useful, low-cost, low-energy, low-noise, and low-pollution vehicle would deprive the poor and middle classes of a most effective means of transport and cause considerable unemployment amongst drivers and associated trades.

Since the tricycle holds exciting promises for both developing and developed countries alike, it seemed worthwhile to rethink the design with a view to maintaining or extending tricycle use. The author's first experience with tricycle building was the design and construction of the cycle rickshaw. This incorporated a simple form of differential gear to allow both rear wheels to be driven but for them to turn at different speeds when rounding a corner; it also showed up the common problems of weakness in the wheels and front forks and lack of braking.

However, it helped to convince Oxfam of the potential for an improved tricycle design and resulted in their financing a technician to

Figure 2-4 Oxtrike chassis

build two prototypes of the new Oxtrike. What follows are some of the major features of the Oxtrike.

(1) The wheels chosen were of 20 inches in diameter, largely for the sake of greater strength. The rear wheels are Raleigh "Chopper" type (20 × 2⅛), the strongest wheels made by that firm. They are especially strong against side load, a requirement which does not arise on a bicycle but is a very real problem on a tricycle. The front wheel (20 × 1¾) and fork are of the carrier-bicycle type and are designed to carry a large load which is forward from the frame over the front wheel.

Figure 2-5 Up-ended view of Oxtrike chassis showing ability to park in a small space, easy inspection of transmission and brakes, and ability to tip out the load. Note Sturmey-Archer three-speed hub gear with double adjustment for primary and secondary chains.

Further advantages of the 20-inch rear wheels are a general lowering of the center of gravity of the load (but giving adequate ground clearance), easier access by elderly or infirm passengers, and a full-width rear seat without an increase in the overall vehicle width, which is only 36 inches. The use of small wheels also reduces the overall length of the tricycle to just over 6 feet 7 inches.

(2) A three-speed gearbox is incorporated into the transmission system; this gearbox is a standard Sturmey-Archer hub gear, a well proved, reliable design. It is used as an intermediate gearbox, as on many motorcycles, with a primary and secondary chain. The sprockets are so chosen that the overall "gear" of the Oxtrike is 31½ inches in bottom gear, 42 inches in middle, and 56 inches in top. These compare with 66½ inches for a normal bicycle with 26-inch wheels and greatly improve the driver's ability to start with a heavy load and to climb at least a slight gradient, thus improving the range and mobility.

(3) Since good brakes are essential for safety, particular attention was paid to this problem; the front brake is the standard pull-rod stirrup type, but the rear brakes are inboard band brakes applied by a foot pedal. This is a powerful and effective method of applying a braking force: each wheel has its own brake drum, mounted at the inboard end of the half-shaft, and the braking effect is applied equally by means of a balance bar. This location of the brakes ensures that they are protected from the rain, essential for safety in wet weather. The brake pedal can be held down by a lever catch to act as a parking brake, a very necessary feature on a tricycle.

(4) Since the construction of a normal bicycle frame is a fairly complex matter, involving thin-walled tubing brazed into special sockets, it is not really suitable for small-scale local manufacture except by the importation of the tubing and sockets. Suitable sockets for tricycle construction may well not

Figure 2-6 Back axle drive of Oxtrike

be available. For this reason the Oxtrike was designed to use mild steel sheet of a standard thickness. This can readily be cut on a foot-powered guillotine, folded on a hand-operated folding machine, and joined by almost any method of welding, brazing, or riveting.

(5) A variety of bodywork types have been designed, including a two-passenger rickshaw with a multiuse body having a hinged tailboard. When lowered, the tailboard allows two passengers to sit facing backward; when raised, it provides for three children to sit facing forward or for the carriage of goods. Larger loads may be carried with the tailboard horizontal.

The chassis can be fitted with a simple open truck body or an enclosed box body for carrying parcels, etc. A hopper body with sloping ends would be suitable for carrying sand, gravel, or other loose material; the size of the hopper would be restricted to prevent

Figure 2-8 Oxtrike with temporary seat

Figure 2-7 Hand-powered machine for making rolling section

Figure 2-9 Neo-Chinese wheelbarrow with 4-foot-diameter wheel of sandwich construction

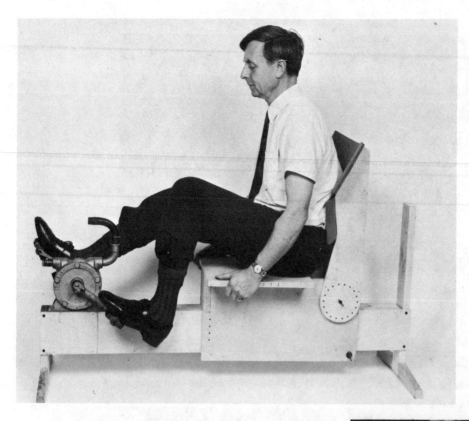

Figure 2-10 Rotary hand pump connected to pedal operation, mounted on fully adjustable sliding-seat (Courtesy of Autometruc Pumps, Ltd.)

overloading. Finally, a 40-gallon drum can be fitted for carrying liquids.

For rougher going, where a four-wheel-drive vehicle such as a Jeep or Land Rover would normally be required, a vehicle is being evolved named the Pedal Rover. It consists of four large-diameter wheels of about 44 inches, each directly pedaled by one of the crew. The front and rear halves of the vehicle are articulated to allow both steering and twisting of the two halves as in modern dumper trucks. It is expected that a payload of 500 to 600 lb. could be transported in such a vehicle.

Even simpler and more versatile is a new-Chinese wheelbarrow, using a single large-diameter wheel under the load rather than the usual small-diameter wheel in front, which leaves too much load on the handles and makes tipping difficult.

Two-wheeled garden carts are available with a 26-inch wheel on either side. These are adequate on smooth ground but tend to yaw badly on rough going; a single wheel cannot be thrown off course and enables

Figure 2-11 Two-man dynapod built by Alex Weir

even the narrowest of footpaths to be used. Rolling resistance is reduced by using a really large diameter wheel. Figure 2-9 shows a 48-inch wheel in a prototype built for the Rev. Geoffrey Howard who later crossed the Sahara, 2000 miles from north to south, pushing a load of up to 350 lbs. singlehanded and averaging over 20 miles per day.

For general garden use, a simple and effective design can be built with a 26-inch diameter bicycle wheel and tire underneath a three-sided tray-type body. Although higher for loading, unloading is easier, especially if a brake can be incorporated to stop the barrow running forward when tipping.

Stationary Pedal Power

Turning to stationary uses for pedal power, three approaches are possible. A hand-cranked device such as a corn mill may be fitted with pedals and arranged with a suitable seat for direct pedaling. Figure 2-10 shows a rotary hand pump of the sliding-vane type mounted on a low wooden trestle or "horse" which is fitted with an adjustable seat; normally only horizontal adjustment is needed for varying lengths of leg. Though not quite as efficient physiologically as the normal saddle above the pedals, this arrangement may prove convenient for many people.

The Dynapod

The second approach is to design a basic stationary pedal-power unit or "dynapod" (from the Greek for "power" and "foot") which can then be hooked up to any device that needs to be driven. A design for such a dynapod was put forward in 1968 but was not built at that time, although Alex Weir from Edinburgh University built one-man and two-man units at Dar-es-Salaam in Tanzania. He

used as a flywheel an old bicycle wheel in which the whole of the space between the spokes was filled with cement. Later Alex Weir made other types of pedal units in Dar-es-Salaam, using square-section tube for the framework. He used them to drive corn grinders and a winnowing machine.

Among the machines which could be pedal driven by means of a dynapod is not only a wide range of hand-driven machines, pumps, corn grinders, forge blowers, grinding machines, etc.—many of which may be better converted to direct pedal drive—but also a great number of machines such as potter's wheels, drilling machines, and horizontal grinding wheels, which are not normally hand driven.

Figure 2-12 shows a commercially available maize sheller which is driven by a very simple but effective form of bolted-on pedal drive unit. It is sold in several countries in Africa and can shell two cobs at a time. A heavy cast-iron fan serves both as flywheel and as a winnower to blow away the husks.

The problems of optimizing pedal power for stationary use are several. The problem of the cyclic torque has been mentioned, but another is the need for a rigid strut between the driving and driven sprockets to take the pull of the chain, which can be roughly twice the rider's weight. There must also be some provision for adjusting the chain. A less obvious need is to keep the rider cool in the absence of forward motion. The Chinese, who have used a simple form of pedal power for hundreds of years, usually provide a roof over the pedalers for shade and to keep off the rain. It might actually be worthwhile to use a small part of the power to drive a fan to keep the rider cool.

The Winch

The third approach to using pedal power is to design the equipment from the start for pedal drive. An example of this is the prototype of a two-man pedal-driven winch

Figure 2-12 Commercial maize sheller (Courtesy of Ransomes Sims & Jefferies Ltd., Ipswich, England)

Figure 2-13 Commercial maize sheller in use in a village in Africa (Courtesy of Ransomes Sims & Jefferies Ltd., Ipswich, England)

shown in Figure 2-15, which is based on the use of two automobile flywheels. The lower one carries the winch drum and is driven by means of the starter gear ring meshing with a starter pinion on a shaft above. This shaft carries two small fixed sprockets, each connected by a chain to a normal bicycle-type chainwheel and pedals, arranged in such a way that the two sets of cranks are at right angles to each other in order to smooth the output.

The shaft also carries a second flywheel which rotates sufficiently fast to store appreciable energy in order to overcome any sudden snag in the load being winched. The gear ring of this second flywheel can be fitted with a pawl to form a ratchet mechanism and prevent the load running backward. For lowering a load the pawl ring may be disengaged and the load lowered under control by means of the braking action of the pedals and also of a caliper-type brake acting on the second flywheel.

The whole unit is mounted on skids so that it can be moved sideways when the pedalers dismount; but when they are pedaling, their weight helps to anchor the winch firmly so that the cable can exert a powerful horizontal pull. Apart from the obvious uses for such a winch in excavations and load lifting, the major use on the land is for cable-cultivation, an old principle in which the motive power for plowing or other cultivation is stationary and only the implement moves across the field. (See the Rodale winch in chapter three.)

Advantages are as follows: a saving in energy, since the motive power—human, animal, or machine—does not have to waste power in moving itself over the soil; avoidance of soil compaction, one of the worst features of using a big tractor; the ability to work even waterlogged ground, as can be seen in China, using an electrically driven winch. For nearly 100 years steam cable plowing was the only mechanized method of agriculture. Small engine-driven winches are

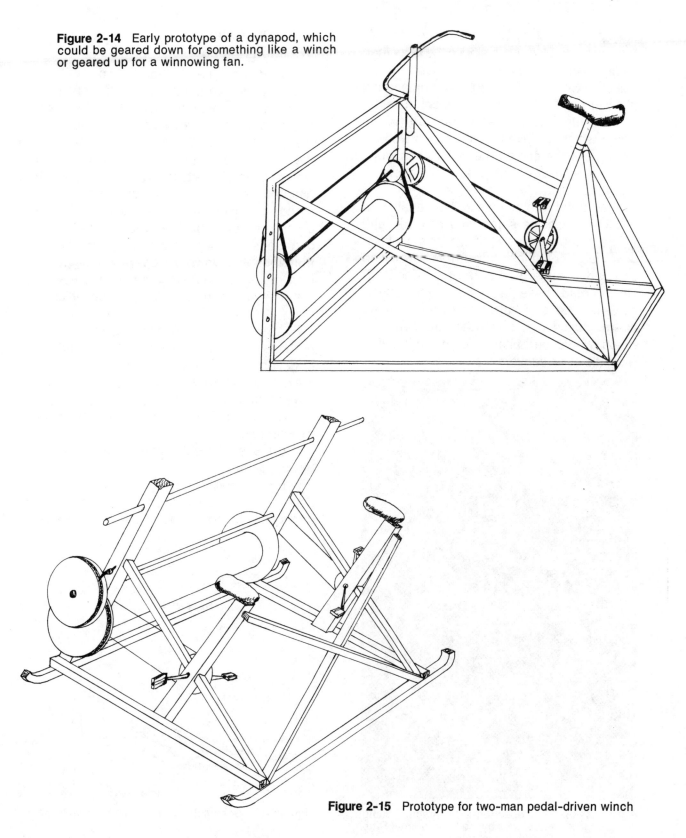

Figure 2-14 Early prototype of a dynapod, which could be geared down for something like a winch or geared up for a winnowing fan.

Figure 2-15 Prototype for two-man pedal-driven winch

used on small steep plots in France and Italy for hauling a plow up a slope, the plow then being dragged down again by hand. At the National College of Agricultural Engineering at Silsoe in Bedfordshire, a recent development is the Snail, an engine-driven mobile winch which is driven along on two wheels, paying out the cable. The winch then stops and hauls in the cable and the process is repeated.

The pedal winch should be capable of tackling much the same type of work, and where manpower is plentiful, as in most less-developed countries, a two-man winch can be used at either end of the plot. The pedalers could have a rest in the shade between spells of hard work! Hand-pushed plows, cultivators, and hoes are available which would be suitable for cable traction with little or no conversion.

Figure 2-16 Foot-operated diaphragm pump developed by the International Rice Research Institute

Pedal Drives for Irrigation Pumps

In Bangladesh and other parts of the Third World a requirement exists for pumping water from a river to the fields. The foot-powered pump developed by engineers at the International Rice Research Institute in the Philippines can lift large quantities of water several feet using only moderate amounts of labor. The operator simply stands on two foot rests at either end of the pump and rocks back and forth. That effort compresses a diaphragm which forces water from the outlet valve. By operating the pump in a rhythmic manner, a continuous flow of water is pumped. This is quite an efficient unit.

Efficient as the bellows pump is, it is perhaps possible to propose a pedal-driven irrigation pump, particularly one that could leave water at considerable heights. Further requirements would be:

(a) low cost but long life with minimum maintenance,

(b) use of local materials or standard bicycle parts,

(c) portability (the pump must not only accommodate to varying river levels but be capable of being moved to different sites as

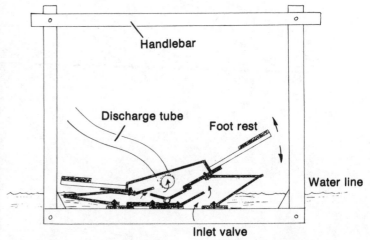

Figure 2-17 Schematic drawing of a bellows pump

required and perhaps dismantled and stored during flood times),

(d) use of pedal power rather than manual operation, since it is two to three times more effective, and

(e) if possible, a two-man operation rather than one, for increased and smoother output as well as for social reasons.

Figure 2-21 shows a proposed design of two very traditional elements in a new way. The pedal unit is of a type used in China for hundreds of years and still in use for a variety of purposes, including low-lift pumping by means of a "square pallet chain pump." This type of pump is not altogether suitable for Bangladesh and similar areas because at

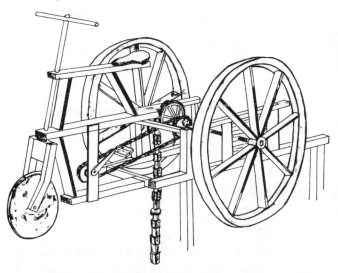

Figure 2-18 Chinese "tricycle" water pump in which wheels also serve as flywheels

Figure 2-20   Close-up of Figure 2-19

Figure 2-19   Chinese wooden water pump used for raising sea water into salt-evaporation beds

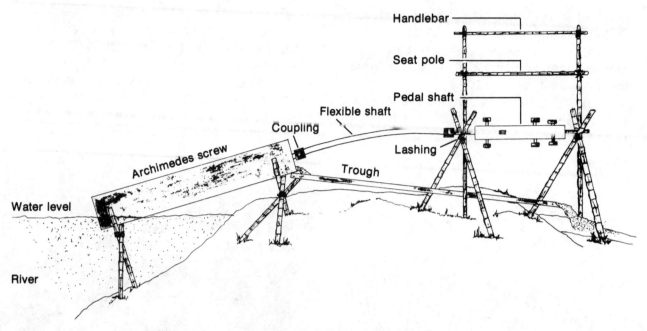

Figure 2-21 Prototype for a pedal-driven Archimedes screw

low-water levels the distance from the bank to the water may be 10 meters or more; also, the construction is complicated and difficult to transfer successfully.

Hence another very traditional device is suggested, the so-called Archimedes screw, which probably originated in Egypt before the time of Archimedes and is still in use both there and in developed countries for certain special purposes. It is said to be up to 80 percent efficient.

One method of constructing an Archimedes screw is to coil up a circular section pipe into a cylindrical helix. It is known that this form was used long ago, but it is not clear what materials were used. A modern version could be made using thin-walled plastic tubing; a particular type has recently been evolved for field drains, in which the tubing is corrugated with a fine pitch to strengthen it and to allow coiling to a small radius. Although this is normally perforated with a multitude of fine holes, if it could be obtained unperforated, it could form the basis of a simple low-cost pump, since the rest of the construction could be done locally.

For example, a stout bamboo could serve as the main axle and the coils of pipe—probably in two-start or three-start thread form—could be held in place by lashing with rope, cord, tape, or any suitable local fiber, using longitudinal strips of bamboo or other wood to form a cage on the outside of the coils. Although the pipe corrugations would give increased friction to the flow of water, their other virtues may outweigh this disadvantage.

The possible dimensions of this type of pump are such that it is almost certainly suitable only for slow speeds, which could enable it to be driven directly from a Chinese two-man pedal unit at up to 30 rpm.

The pump is connected to a pedal unit situated near the top of the river bank. Since the maximum slope of an Archimedes screw of any type is about 30 degrees, a length of 20 feet (6 meters) is needed and should be possible with all types. This leaves a gap of 13 feet (4 meters) or so to be bridged to the horizontal shaft carrying the pedals. A simple and effective method is to use a steel rod of such a diameter that it will transmit the torque but will bend over its length to accommodate the 30-degree difference in slope between pump and pedal unit. A suitable diameter is about ¾ inch (18 mm). By making the pedal shaft with a full-length axial hole to accommodate the rod and providing a coupling at one end which can clamp the rod firmly, e.g., by use of two cotter pins of cycle type, adjustment to length can readily be made. The best method may be to insert the rod alone into the coupling on the pump, then to thread the pedal shaft over the upper end of the rod, then pull down the pedal shaft onto its supports, finally clamping the rod at its upper end.

These supports may conveniently be tripods made from three bamboo poles, one of which is larger and extends upward to help support a stout horizontal pole for use as a seat and an upper horizontal pole acting as a handlebar. The three poles forming a tripod are lashed firmly together, leaving short extensions to provide a three-point support for a spherical wooden bearing, drilled across a diameter to accommodate the shaft of the pedal unit. It may be necessary to hold down the bearing to its tripod, which could be done by use of a metal ring on top of the housing, the ring being then lashed to each of the three poles. Such an arrangement would give a self-aligning bearing; the diametrical hole may be lined with a brass or plastic tube to provide a better bearing surface if the particular wood used is not adequate, but in most cases wood should suffice. A method used by Alex Weir is to boil the wood in oil (e.g., old engine oil) for 12 hours to provide built-in lubrication.

Pedal Drives for Borehole Pumps

There is widespread need in India, Pakistan, Bangladesh, and many other countries for a reliable form of pump which is capable of lifting water from depths of from 20 to 330 feet (6 to 100 meters) or more. The normal pump used for such purposes is a well proved design and gives good reliable service in engine-driven and windpump installations. There is evidence of widespread failure in hand-operated village use; therefore, it seems worthwhile evolving a pedal-driven version on the grounds that it is up to three times as effective as hand operation, enabling greater quantities or greater depths to be achieved with minimum energy expenditure.

Figure 2-22 shows an arrangement based on a traditional Chinese method of pedaling and used for hundreds of years for operating pumps, windlasses, etc. A horizontal wooden axle is fitted with two sets of pedals, each set consisting of four arms spaced 90 degrees apart with a short cylindrical pedal at its outer end. The two sets of pedals are arranged at an angle of 45 degrees to even out the pedaling torque. The shaft has bearings at either end supported by crossed poles. One pole at either end extends upward to cross a second horizontal pole at a convenient height to serve as a seat for the two pedalers, who may also hold on to a third horizontal pole in front of them and lean backward against a fourth pole.

The main pedal shaft is extended at either end by means of a steel shaft which passes through the bearing and overhangs a short distance at each end. A standard bicycle left-hand crank and pedal is fitted at one end and at the other is fitted a standard chainwheel and crank or a wooden pulley with a diameter of 12 inches or more.

The left-hand pedal is connected to a wire cable or a rope which passes upward and over a bicycle wheel used, without a tire, as a pulley wheel. The wheel is supported by the rear part of a standard bicycle frame suspended from the two upper horizontal poles. The other end of the cable is connected to the upper end of the pump rod; a leather strap may be used to line the wheel rim to prevent damage to rim or cable. The strap can be joined by a thong at the bottom of the

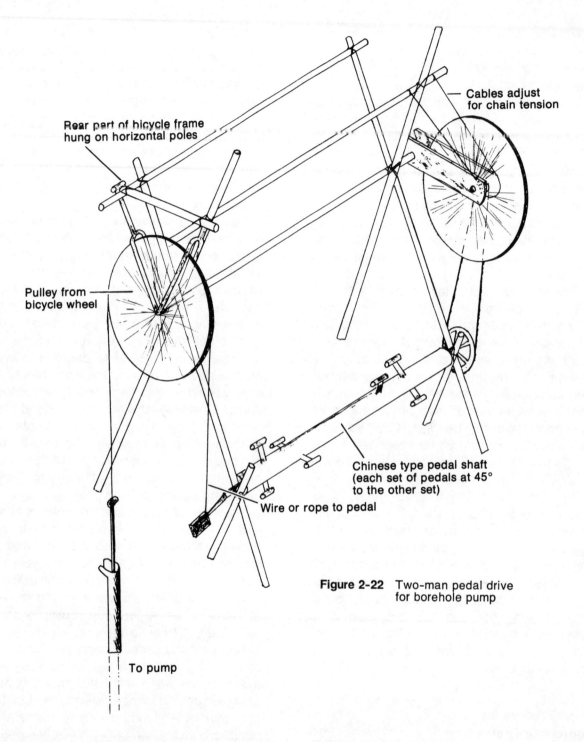

Cables adjust
for chain tension

Rear part of bicycle frame
hung on horizontal poles

Pulley from
bicycle wheel

Chinese type pedal shaft
(each set of pedals at 45°
to the other set)

Wire or rope to pedal

Figure 2-22 Two-man pedal drive
for borehole pump

To pump

wheel, since the wheel will reciprocate through an angle of only about 60 degrees.

The chainwheel or pulley at the other end of the pedal shaft drives a flywheel at a higher speed by means of a step-up drive; using a standard 46-tooth chainwheel and the smallest standard fixed sprocket of 15 teeth, a ratio of about 3 to 1 is available, giving a flywheel speed of about 90 rpm for a pedaling speed of about 30 rpm (half the normal bicycle pedaling speed, since there are four pedals per revolution). With a 12-inch pulley and a flat leather belt drive to a pulley of 2- to 3-inch diameter, a ratio of 4:1 to 6:1 could be obtained, giving a higher flywheel speed.

The flywheel itself could be made from an old bicycle wheel with the space between rim and hub filled with cement and/or the rim wound with whatever type of wire is available. The wheel is supported from the upper horizontal poles in such a way that the tension in the chain or belt can be adjusted. A possible method is to use a front fork pivoted to one pole and a "Spanish windlass" (a double rope twisted by a piece of wood to tighten it) attached between the other pole and a short yoke connected to either end of the axle. Such an arrangement is suitable for chain drive but a different arrangement is needed for belt drive, since a longer axle is required. Possibly a wooden fork could replace the bicycle fork. The use of a flywheel should considerably improve the smooth operation of the pump by helping to lift the pump during the operating stroke.

Some improvement may be achieved also by partially counterbalancing the weight of the pump and operating rod; the counterbalance could be attached below the crank pedal, between the pulley wheel and the circle swept by the crank, or to one side of the pulley wheel.

The pump as described should have the following performance with two men pedaling at 30 rpm:

Pump bore, inches	1½	2	2½	3	4
Lift feet	204	108	75	51	27
Delivery, gallons/hour	135	240	373	540	960

Figure 2-23 shows an alternative arrangement for one man pedaling in the normal bicycle mode. There is a T-shaped wooden base carrying a tripod on which is mounted the saddle and the pedals. From the chain-wheel the chain is taken forward to drive a bicycle rear wheel, modified to form a flywheel by pouring cement into the space between hub and rim. The freewheel is retained but should have a 22-tooth sprocket instead of the normal 18-tooth sprocket. Such larger sprockets are obtained and are used, for example, in Bangladesh on cycle rickshaws. On the other side of the wheel is fixed a 15-tooth sprocket—this may be the most difficult part of the construction, but some rear wheels do have threads on both sides of the hub; with other wheels some other method of attachment could probably be devised.

A second chain conveys the drive from the 15-tooth sprocket to a second chainwheel and pedals vertically above, giving a speed of about 45 strokes/min when pedaled at 66 rpm, with a flywheel speed of about 140 rpm. The second chainwheel carries a normal crank and pedal to which is connected a rope or wire cable running over a bicycle wheel used as a pulley in the same way as described for the first pedal unit except that here the wheel may be supported on two posts and braced to the pillar supporting the flywheel and second chainwheel by means of a strut. The tension of the first chain is resisted by a compression member consisting of a bicycle front fork. The threaded top portion of the fork enters a hole in the saddle post and a threaded nut may be used to adjust the tension in the chain. The tension in the second chain may be adjusted by means of a screw which raises one end of the horizontal member carrying the pedal shaft, the front end of which is hinged to a lower member mounted on top of the left-handed post. The upper member may also be used to carry a wooden bar or metal tube for use as a handlebar.

The upper pedal shaft may also be fitted with a normal left-hand crank and pedal so that both it and the right-hand crank may be used as handles to assist the pumping effort by an extra person on either side. A balance

53

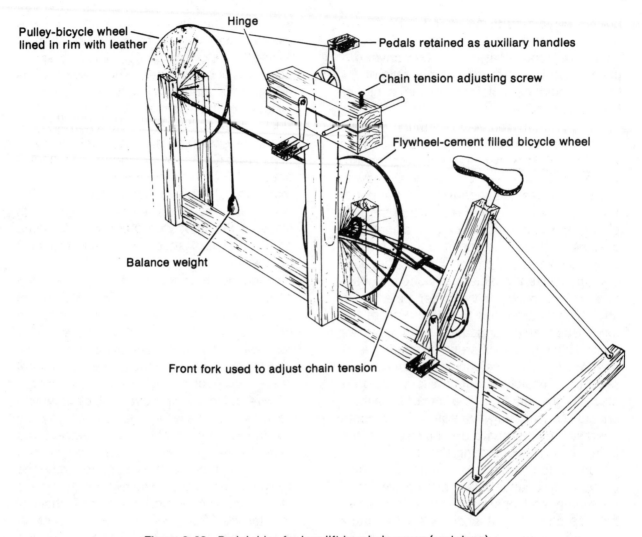

Pulley-bicycle wheel
lined in rim with leather

Hinge

Pedals retained as auxiliary handles

Chain tension adjusting screw

Flywheel-cement filled bicycle wheel

Balance weight

Front fork used to adjust chain tension

Figure 2-23   Pedal drive for low-lift borehole pump (prototype)

weight may be hung from the pulley wheel to reduce the dead weight of the pump and its operating rod.

Figure 2-24 shows a further arrangement which is better suited to deeper boreholes. The stroke is shortened, to as little as 5 inches, and the speed reduced to about 20 rpm. These changes are effected by using an old automobile flywheel with its starter gear ring meshing with a starter motor pinion to give a large reduction of about 13:1; the pinion is driven by a sprocket-and-chain drive from a standard chainwheel and ped-

als to a 15-tooth sprocket on the same shaft as the pinion. This shaft also carries a cement-filled bicycle wheel to act as a flywheel.

There are almost certainly many other arrangements possible for pedal-driven borehole pumps, as well as hand-driven variations, but the ones described here are probably worth trying as they appear to offer a solution to the main problems of operation and can be built readily from local materials or easily obtainable parts. The bearings throughout are standard bicycle hub or pedal shaft bearings; in some cases a 3-inch

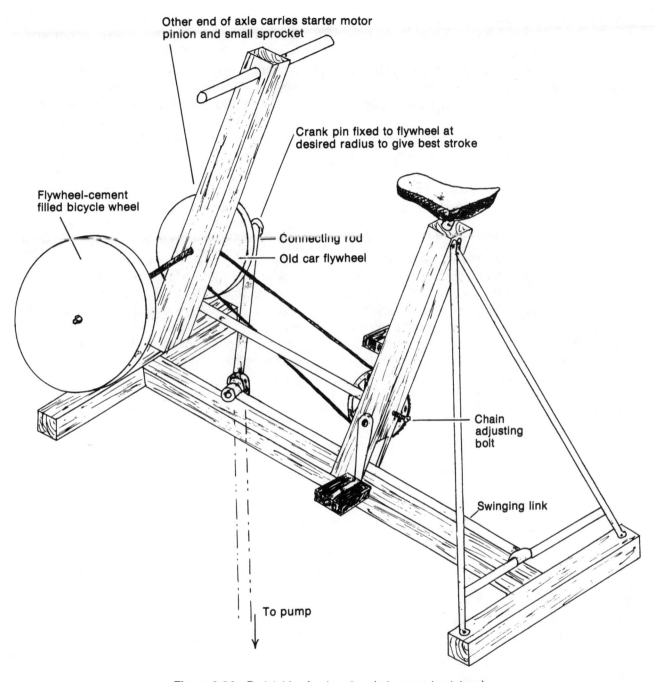

Other end of axle carries starter motor pinion and small sprocket

Crank pin fixed to flywheel at desired radius to give best stroke

Flywheel-cement filled bicycle wheel

Connecting rod

Old car flywheel

Chain adjusting bolt

Swinging link

To pump

Figure 2-24 Pedal drive for deep borehole pump (prototype)

length of 1½-inch diameter tube is used, threaded internally with left-hand and right-hand threads to take the ball race cups. These tubes may be obtainable as standard parts or made by a small manufacturer possessing the necessary taps. A wooden member, e.g., 3 inches square, may be drilled with a 1½-inch diameter hole to accommodate the tube, or the tube can be welded or brazed to a plate bolted or screwed to the member.

To reiterate the main advantages of these forms of pump drive:

(1) A fixed pump stroke should give a long, trouble-free life.

(2) Pedal operation is much more effective than hand operation, especially with a flywheel.

(3) Bicycle parts, as well as the required constructional and maintenance skills, are widespread.

A major obstacle to this type of simple design has been the lack of any unit dedicated to the design and testing of prototypes. There is little incentive for commercial firms to undertake such work, since there is no obvious financial return. Not insignificantly, the whole idea of solving real problems by simple means—rather than by computers— is foreign to the present ideas of academic work. However, the climate of opinion is changing, and there are several universities throughout the world where simple technologies are being taken seriously. A notable example is the ASTRA cell at Bangalore in India (Application of Science and Technology to Rural Areas). And it is likely that we will establish a Simple Technology Development Unit at Oxford University.

In the meantime, encouragement and good work is coming from America. A National Center for Appropriate Technology, recently funded by the Congress, will, it is hoped direct some of its energy toward pedal power in work and transportation.

And there is Rodale Press, Inc., which is vitally concerned about all aspects of appropriate technology, including pedal power. In the future perhaps we will see pedal power as one of our common bonds. The majesty and simplicity of the bicycle has applications for all of us.

WHEEL

Circulate the wheel
collecting gossip
from Archimedes
and the man on the bellow's pump,
cut a message on a band saw
fashion legs on a lathe
draw water and life
from the well.

James C. McCullagh

CHAPTER THREE
Multiuse Energy Cycle: Foot-Powered Generator*

By Diana Branch

During the last two decades E. F. Schumacher, the noted economist, has popularized the idea of people-oriented technologies, of intermediate technologies which would be small, inexpensive, and relatively simple. In calling for such technologies, particularly for developing countries, Schumacher argues that we must devise a full range of tools, devices, and techniques to occupy that sensitive middle ground between primitive plow and combine.

In some quarters intermediate technology has been advanced as one way in which developing countries can control their own fate. On the other hand, because this approach takes under close scrutiny the people and ecology of a region, it has great appeal to Americans, many of whom are fleeing the traditional economics where electric gadgets open and close the day.

America, because of a general availability of machine tools and generous resources,

has long been a land of tinkerers and inventors, as the Patent Office will readily affirm. The workshop has occupied a central position in the American home for generations.

Thus, Schumacher's call for human technologies and systems has reached a receptive audience in America. The audience, served by *Appropriate Technology, RAIN, Alternate Sources of Energy, Co-Evolution Quarterly, Mother Earth News, Organic Gardening and Farming,* and other publications, has discovered that intermediate technologies provide them with a rationale for controlling their fate, with a life-style worth pursuing.

The search in numerous countries is for tools and machines that are human centered. Not surprisingly, the bicycle, that quintessential human machine, is once more gaining in popularity. It is certainly the object of serious consideration in America. It is once again an item in transportation and

*Energy Cycle is a registered trademark owned and used exclusively by Rodale Resources Division.

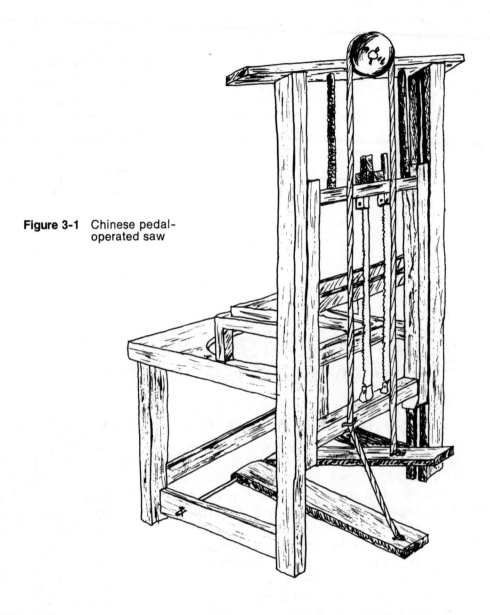

Figure 3-1 Chinese pedal-operated saw

recreation. Ingenious Americans, learning from nineteenth century inventions, are discovering ways to adapt the bicycle to railroad tracks.

Americans, in a quest for simplicity, are building pedal- and treadle-operated saws and lathes for the home workshop. Technologies from Asia and Africa are being embraced.

Mindful of these developments, Rodale Press, Inc., which has long had an abiding interest in "people power," has given considerable thought to the bicycle's position in twentieth century American life.

The Energy Cycle
Genesis of an Idea

In this climate of bikology, Rodale's Research and Development Department contemplated ways in which to put the principle of the bicycle to maximum use for work purposes. Researchers knew that the muscle energy conversion of the bicycle is around 95 percent. If the bicycle could be so efficient in transportation, what would happen if the same efficiency was delivered to work situations? In their huddle of creativity, R & D personnel sought to find the answer.

With certain key design criteria in mind (simplicity, ease of operation and maintenance, and low cost) and spurred on by Robert Rodale, inventor Dick Ott worked

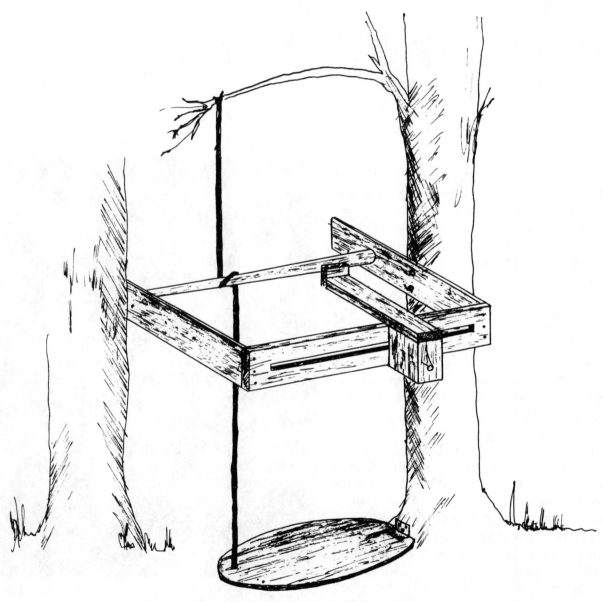

Figure 3-2   Wood lathe (Arkansas, 1977)

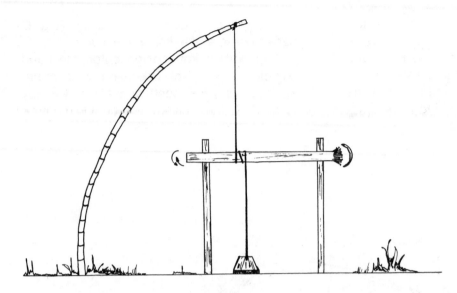

Figure 3-3 Coconut shredder (Africa)

on a prototype. He stripped an old bicycle down to its frame, mounted a power head over the pedals where the seat had been, then ran a pulley system to the sprocket. He put the seat from an old typing chair in place of the handlebars, and the Energy Cycle was born.

The principle of the Energy Cycle foot-powered generator was uncomplicated. The energy generated by pedaling is transmitted to a power takeoff head. A conventional chain-and-sprocket is used as the drive chain. The input shaft of the machine or tool to be powered is coupled directly to the power takeoff shaft.

From the beginning the designer attempted to avoid the disadvantages associated with rear power takeoff assemblies—which he did, by bringing the chain drive up to the front of the frame where the operator could work with his hands.

Researchers found that this prototype could accommodate a number of attachable tools, including an egg beater, can opener, nut chopper, food grinder, and fish skinner. The results were encouraging. The designer felt, at least theoretically, that the Energy

Figure 3-4 Dick Ott's first attempt at harnessing the power-of-the-pedal, which he affectionately called "Pedal Pusher"

Cycle generator could be useful in numerous tasks now performed by handcranks and small horsepower motors.

Accordingly, R & D upgraded the first design. For convenience, a large work table was added to the unit, which enabled the operator to perform numerous tasks without leaving his seat. Machinist LaMar Laubach replaced the original frame with one of 1¼-inch pipe, which added sturdiness and allowed the operator to bear down more

easily for power. He also built the seat a little lower and farther back from the pedals than on the original frame. This modification put the rider's legs closer to the horizontal, making it easier for him to work at strength through a full rotation. The frame rested on 30-inch lengths of pipe, improving stability.

Workers added a larger sprocket to the cycle for more power. A fifth pulley increased the speed. An idler took the slack out of the belt.

Testing Program[*]

Buoyed by hundreds of suggestions from people who had read about the unit, investigators tested the versatility of the machine, which they found to be considerable. Kitchen aids which the unit powered during early trials include: food grinder, food shredder,

Figure 3-5 Streamlining of the cycle frame to cut down materials, weight, and width without loss of stability

Figure 3-6 Energy Cycle powering a corn sheller

*All Energy Cycle foot-powered generator test results described throughout this chapter, whether general or specific, quantified or not, are only tentative and are merely representative of tests up to the time of the preparation of this chapter. Such tests are still continuing. Field and actual results may vary from these tentative test results.

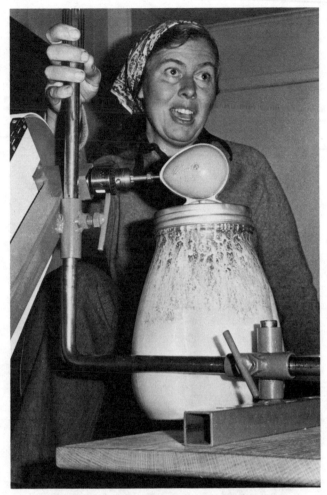

Figure 3-7 Churning goat butter

Because hand grinding is difficult and electric mills expensive, grinding grain is a popular application of pedal power. R & D personnel found that they were able to grind 5 lbs. of wheat in 20 minutes (an electric mill would take about 14 minutes; hand grinding, well over an hour).

Some of the kitchen tasks tried under test conditions include pitting cherries, French-slicing beans, slicing meats and cheeses, milling vegetables, grinding hamburgers, and pureeing fruits and vegetables. Some jobs were made easier; others faster. With some, there was no advantage in using the Energy Cycle. Researchers report that when working with cherries, a person can sort, pluck, and feed with the hands while the feet

can opener, dough kneader, batter beater, butter churn, ice cream freezer, flour mill, cherry pitter, apple peeler and corer, potato peeler, knife sharpener, juicer, sweet corn kernel cutter, dried corn sheller, meat slicer, sausage machine, fish skinner, etc.

The farm machinery applications include the irrigation water pump, feather plucker, cultivator, weeder, harrow, discer, plow, potato digger, corn sheller, grain cleaner, rice polisher, and oatmeal roller.

Still other pedal-tested tools include a wheel grinder, stone polisher and buffer, drill, jeweler's lathe, wood carver, potter's wheel, and battery charger. If there are wheels or cogs in the machine, chances are that the Energy Cycle can power it.

Figure 3-8 Cherry pitting

Figure 3-9 Making applesauce

Figure 3-10 Potter's wheel powered by
Energy Cycle

Figure 3-11 Jewelry lathe attached
to Energy Cycle

Figure 3-12 Pumping water

do the pitting. Time spent in slicing cabbages and beans can be cut in half. On the other hand, pureeing vegetables is hardly worth the effort of setting up the machinery unless quantities are involved.

In another test of pedal power, workers powered a 3,500-gallon-per-hour water pump to capacity, sending water streaming out 30 feet through the hose nozzle. By comparison, the average garden hose puts out about 420 gallons an hour, depending on water pressure.

Under actual farm working conditions test units of the Energy Cycle generator performed numerous tasks. In a field left fallow for a year, the cycle pulled a plow through the grass- and weed-covered soil, much as a farm horse might have done at the turn of the century. In this two-man operation, one pedaled the winch that drew the plow through the soil while the other guided the plow. It took the two approximately an hour to plow 1,500 square feet. Similarly, the winch apparatus cleared a weed-infested bean field with relative ease. Workers realized that the pedal winch, which is capable of delivering more than a thousand pounds of torque, had a tendency to disable ordinary hand tools. For that reason Rodale

Figure 3-13 Early attempt to use pedal power in the garden

Research and Development is designing special tools to be used with the Energy Cycle (see the winch section which follows).

A number of quantified tests indicate that the cycle is capable of electrical output. A simple kit—auto generator, 12-volt battery, and inverter—makes it possible to generate and store electricity for possible use in the home. Some tests have shown that, when the unit is hooked up to a television, 20 minutes of 70 rpm cycling gives 30 minutes of viewing time. It is possible to use an inverter that converts the power stored in a 12-volt car battery into the 110 volts needed to operate standard electrical appliances. However, researchers feel that in the long run it would be more advisable to use appliances that are run on direct current from the battery, such as those found in campers and trailers. Stereos and tape decks are good things to run off a battery as they draw low voltage.

Refinements in Design

Convinced of the worth and versatility of the Energy Cycle, the designer streamlined the cycle frame by cutting down on materials, weight, and width (18 inches vs. 30 inches, without the table). The bent pipe frame ac-

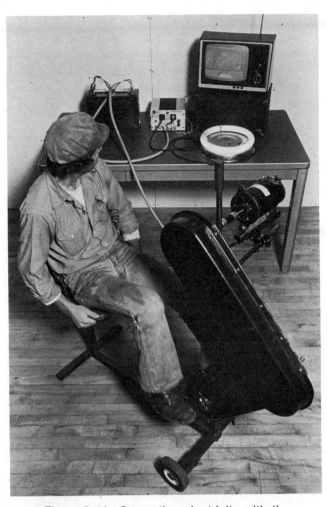

Figure 3-14 Generating electricity with the Energy Cycle

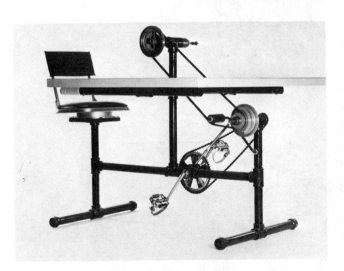

Figure 3-15 Prototype No. 2

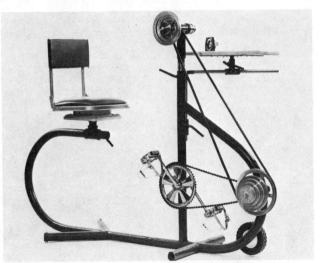

Figure 3-16 Prototype No. 3

commodated a sliding seat. A small wheel on the front of the cycle frame allowed the unit to be tilted and wheeled. By using quick-locking bolts, the designer carried the adjustable features over into both the machine head and the table. A flywheel was added to reduce unevenness.

One of the biggest problems encountered by the researchers was to find a universal means of attaching each implement. Most of the tools used had handles which could be unscrewed, leaving a threaded shaft to work with. They tried clamping the Jacob's chuck right down on the threaded shaft, but the grip was not secure. Pedaling wore down the thread.

Next they tried putting an extension on that shaft. Cutting the head off a hex bolt, they unscrewed it half way onto a nut and then screwed the nut onto the tool's threaded drive shaft. The extension end of the hex bolt was unthreaded, giving the chuck something to be clamped to. As long as the drive shaft and the connection to the cycle were perfectly aligned along a horizontal axis, everything worked well. Occasionally, the stress was too much for the bolt and it snapped off. After that they made sure that only hard-steel bolts were used.

As a further refinement they designed a power adapter to replace the chuck. The adapter transferred the driving force to the handle rather than to a small bolt. This technique worked well but meant that the handle had to be cut off, rendering the tool inoperable by hand. Researchers continue to explore the most effective ways to attach implements to the Energy Cycle generator.

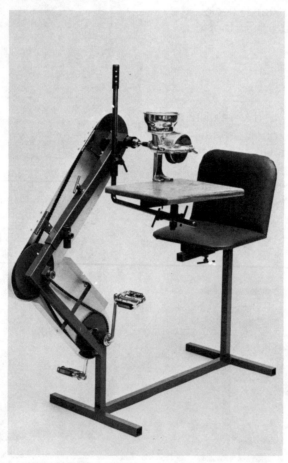

Figure 3-17 Production model Energy Cycle

Winch

Rodale's Research and Development Department believes that the Energy Cycle foot-powered generator is an extremely versatile pedal unit for use around the home, garden, and farm. Moreover, researchers feel that the cycle has application, not only in America, but also in other developed and developing countries.

In fact, because the Energy Cycle held so much promise in the garden and on the small farm, the research team built a stationary pedal-power winch—a specialized winch for pulling.

Economical to build, it can pull up to 1,000 lbs. or more, amplifying human power almost 10 times. It can pull a snow plow, dislodge small stumps, or serve as a power center in the garden to pull seeders, cultivators, harrows, and hay rakes.

Finding tools to use with the winch might be a bit of a problem. Operating hand tools with the added power of the winch can cause them to bend or break. They should be treated carefully or reinforced. Modifying garden tractor utensils to make them lighter and less ambitious in the volume of work they attempt to accomplish at once holds considerable promise. But ultimately, if the pedal-power winch is to ever become a basic

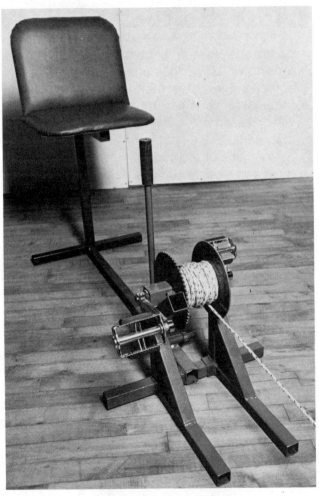

Figure 3-18 Pedal winch

Figure 3-19 The winch assembly is built into a frame which also supports the seat.

tool of society, implements will have to be specially designed for it. Rodale researchers have already tackled the challenge.

The design of the winch itself is quite simple. The unit is basically two pedals separated by a spool mounted on bearings. The pedals serve as a direct drive to the spool upon which the towing cable winds. A brake serves as a kind of a ratchet, holding the cable taut.

The brake handle, made from a length of steel stock with a 90-degree bend at one end, rests inside a piece of steel tubing and is secured in place on either end of the bend by bearings. A stop is welded to one bearing so the brake does not fall back toward the

Figure 3-21 The winch is a two-man operation.

pedaler. Welded on the handle is a metal extension which wedges against the tooth of a sprocket (one side of the spool) from the force of a spring to brake the spool. The winch assembly is built into a frame which also supports the seat.

Most jobs attempted with the winch will require two people, but as designer Dick Ott put it, "No one likes to work alone anyway. This gives the family a chance to develop a working relationship and to establish a family fun center in the garden."

Beyond any philosophical advantage, however, is the new dimension the unit offers in gardening—large-scale intensive gardening. Using a conventional garden tractor, one is limited to 24-inch spaces between rows to conform to the tractor's wheel spacing. Doubling the productivity of your lot, the winch will perform gardening tasks on 12-inch centers.

And closer rows usually mean less weeding. As plants grow thick, less sun penetrates the row spaces and weed growth is inhibited. No longer limited by the height of your tractor, you can weed longer into the season around taller plants. Especially advanta-

Figure 3-20 Close-up of brake handle and metal extension that wedges against the teeth of the sprocket to brake the spool

Figure 3-22 A specially designed, multiuse frame with a cultivator attachment to harness the full potential of the winch

Figure 3-24 The winch performs gardening tasks on 12-inch centers rather than 24-inch, thereby increasing the productivity of the garden.

Figure 3-23 Frame with plow attachment

geous for the organic gardener, the winch helps relieve some of the temptation to use sprays for weed control.

Those at Rodale Press, Inc., who have been involved with the evolution of the Energy Cycle generator and the winch believe these versatile tools partly satisfy Schumacher's admonition for intermediate technologies, for ways in which people can take charge of their own fate. Tools such as this can do much to encourage efficiency and self-sufficiency at home and abroad.

And in order to help make the above goals a reality for many, building instructions are provided for a pedal-powered foot generator based generally on the present Energy Cycle model, as well as for a rear-wheel bicycle adapter.

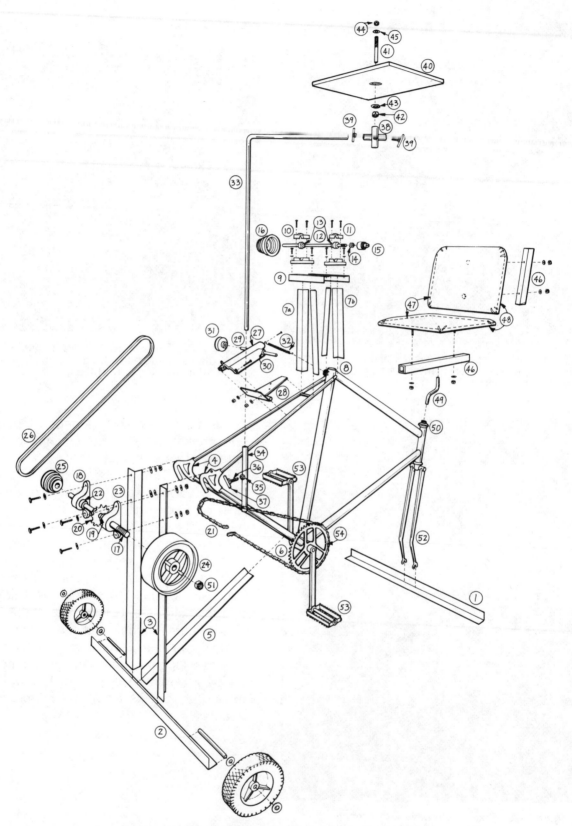

Figure 3-25 Exploded view of homemade Energy Cycle

Homemade Foot-Powered Generator
Materials
FRAME
   Bicycle frame with:
      front[52]* and rear[4] forks
      pedals[53]
      pedal crank[54]
      chain[21]
   The following lengths of 1″ angle iron (approximate, measure as you go)
      5   2′ lengths[1 2 3 5]
      2   10″ lengths[7a]
      2   12″ lengths[7b]
      2   6″ lengths[9]

DRIVE ASSEMBLY
   4   ½″ bore self-centering pillow blocks[10 11 18 23]
   4   ½″ bore bushings[13 19 22]
   2   ½″ bore step sheaves (3 or 4 steps)[16 25]
   1   ½″ bore 10- or 12- tooth sprocket (bicycle sprocket)[20]
   a   ½″ 20-thread Jacob's chuck[15]
   V-belt (appropriate length)[26]
   a 12″ length of ½″ dia. steel stock (½″-20 right-hand thread on one end—1½″)[12]
   a 14″ length of ½″ dia. steel stock (½″-20 left-hand thread on one end—1½″)[17]
   1   ½″-20 right-hand thread nut[14]
   1   ½″-20 left-hand thread nut[51]
   8   ¼″ nuts, bolts, and washers for the pillow blocks[55]
   extra chain links (if needed)
   flywheel—½″ bore (optional)[24]
   toe clips (optional)

IDLER
   One-piece grinder shaft assembly[27] complete with:
      bronze bushings[30]
      8″ length steel stock same dia. of bushings[29]
      2″ dia. pulley to fit shaft[31]
   large gate hinge[28]
   #62 spring[32]

TABLE
   a 3′ length of ¾″ steel stock[33]
   a 16″ × 11″ × ¾″ hardwood board[40]
   the following lengths of ¾″ ID steel tubing
      1   6″ length [34]
      2   3″ lengths[38]

*Numbers refer to parts labeled on photographs.

a 6″ length of ¾″ steel stock threaded 1½″ on one end[41]

3   ⅜″ nuts[35]

3   ⅜″ bolts[36]

3   2½″ lengths of ¼″ steel stock[37]

2   ¾″ nuts[44] [42]

2   ¾″ washers [43] [45]

SEAT

2   12″ lengths of 1¼″ square metal tubing[46]

2   12″ × 15″ × ¼″ pieces of plywood[47]

2   12″ × 15″ × ¼″ foam rubber[48]

2   15″ × 18″ pieces of vinyl cloth

1   8″ length of ⅞″ dia. steel shaft[49]

Building Instructions (Based generally on the present model of the Energy Cycle foot-powered generator)

Use these general guidelines of this model to adapt to the materials you have available. Read the instructions thoroughly, following the photographs before proceeding with construction. Be sure you understand the directions before starting and take time to improve the design to best suit your situation and materials.

Tools for building this unit should be found in many common workshops. To assemble the frame you will need to use either welding equipment or an acetylene outfit. That is the part of construction requiring some expertise. Other tools include a hacksaw, drill, wrenches, allen wrenches, clamps, file, and pliers.

The basic frame of the cycle should be easy to scrounge. Any old bike frame will do. You won't need wheels, tires, or handlebars but be sure to find a frame including the front and rear forks, pedals, crank, and chain. All new frame support pieces will be fashioned from 1-inch angle iron.

Cut a 2-foot length[1]* of angle iron and tack weld it horizontally to the bottom of the front fork[52] of the bicycle to form the back end of the Energy Cycle. (Do not be confused. In building the cycle, the bike frame is turned around so that the back of the bicycle becomes the front of the unit.) Build a T-frame support for the front end from one 2-foot length[2] of angle iron (to rest on the floor) and two vertical pieces.[3] Welded to the horizontal support and the back fork[4] of the bicycle, the vertical pieces should be long enough to keep the pedals at least 4 inches off the floor. For extra support, weld a piece of angle iron[5] between the front horizontal support and the crank section[6] of the frame.

Next, you will build a power-head support where the seat used to be. Cut four lengths[7] of angle iron to extend vertically from the front fork to a height 38 inches off the floor. Position one post on either side of the front fork, close to the former bicycle seat connection.[8] A platform[9] to support two pillow blocks will rest on top of the four vertical posts; therefore, the width of the pillow blocks will determine the position of the second set of posts (3 or 4 inches forward on the frame). Clamp, making sure all four posts are level, and weld.

Cut two 6-inch lengths[9] of angle iron and weld one horizontally on top of each set of support posts[7] to form the platform. Mark and drill holes and bolt two ½-inch self-centering pillow blocks[10,11] in place. Next, insert the threaded end (right-handed thread 1½ inches up shaft) of a ½-inch × 12-inch

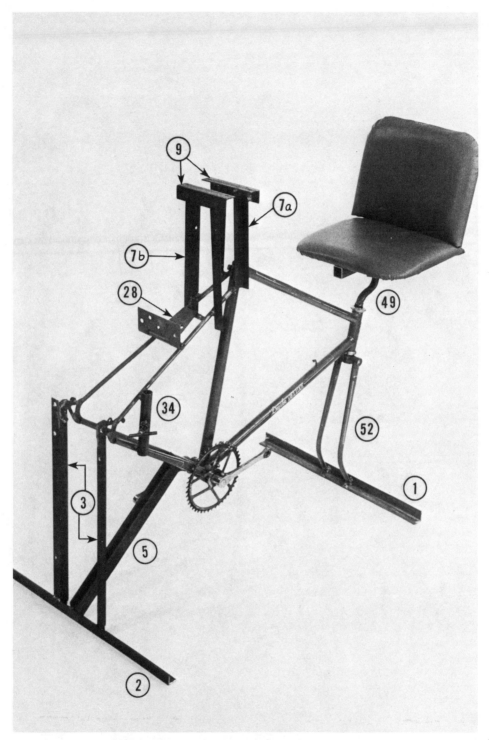

Figure 3-26 Basic frame with seat, power-head support, hinge for idler, table support tubing, and pedal crank

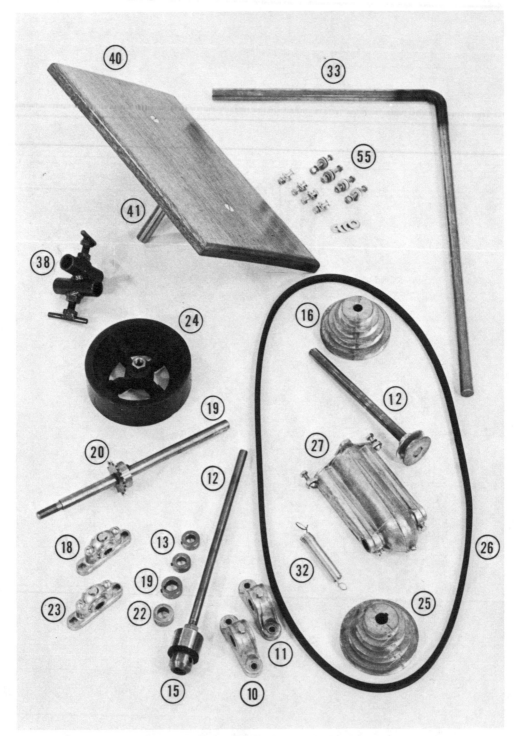

Figure 3-27   To help keep costs down, it should be easy to scrounge
many of these parts.

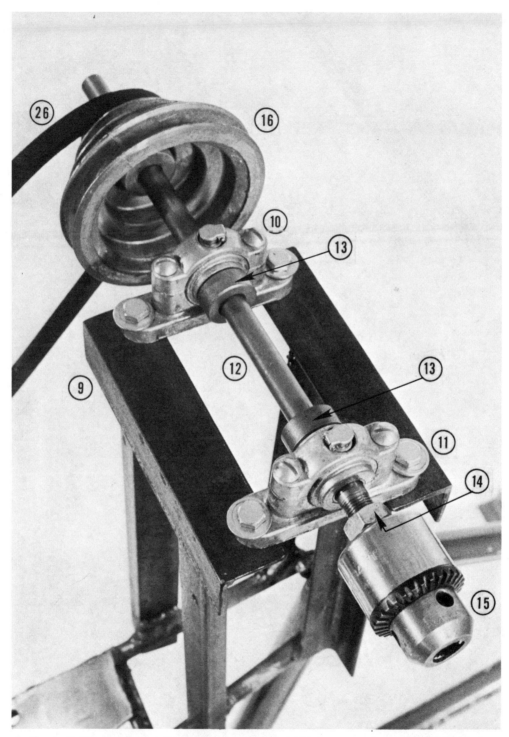

Figure 3-28 Power head: self-centering pillow blocks support an axle which transfers power from the V-belt to the chuck.

Figure 3-29 The table support bar is supported by a 6-inch length of tubing and locked in place by a hand-tightened "T-bolt."

Figure 3-30 The table, which can be moved to most positions, offers a universal means of attaching various tools to the cycle.

steel shaft through the right pillow block.[10] Place two ½-inch bushings on the shaft[12] and push the shaft through the other pillow block.[11] Screw a ½-inch nut[14] three-quarters of the way onto the thread and screw the Jacob's chuck[15] onto the shaft until firmly wedged against the nut. Slide the shaft to the right so that the chuck is as close to the left pillow block[11] as possible without touching. Place a step sheave[16] on the right end of the shaft. Do not secure the fittings yet.

At the front of the cycle, at the top of the T-frame support, drill holes for and bolt the two pillow blocks in place. Insert the threaded end of a 14-inch shaft[17] through the right pillow block.[18] In order, place a bushing,[19]

sprocket,[20] chain,[21] and another bushing[22] onto the shaft and slide the shaft through the other pillow block.[23] The threaded end is for a removable flywheel[24] and the right end for the other step sheave.[25] If the top sheave[16] has been mounted with the small pulley on the inside, be sure the large sheave[25] is on the inside on the bottom shaft.[17]

Check to see if the chain fits the new sprocket.[20] If it does not, remove the master link (slightly larger link) and either add or remove links until the chain is taut. With the chain in place, pedal forward a few revolutions so the front sprocket[20] can align itself. Then align all the bushings [13, 19, 22] and sheaves [16, 25] and file "flats" onto the shafts

77

Figure 3-31 Research and Development personnel adapted a one-piece grinder shaft assembly to perform the task of an idler—a mechanism to remove slack from the V-belt.

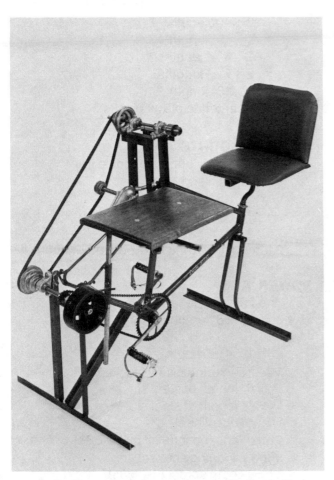

Figure 3-32 Homemade Energy Cycle built by Rodale R & D for less than $60

where the allen screws line up so they will have a flat surface to tighten to. Oil all pillow blocks.[10, 11, 18, 23]

Select a V-belt[26] to fit loosely on the step sheaves to make changing gears a simple task. An idler to remove the slack can be easily made from a one-piece grinder shaft assembly[27] mounted on a gate hinge.[28] First, weld a large gate hinge[28] to the frame just below the power-head support. The hinged end should extend toward the front of the bike and be able to open a full 180 degrees on a horizontal axis. Bolt the one end of the assembly to the flexible part of the hinge and insert an 8-inch shaft[29] through the bronze

bushings[30] at the other end. Add a 2-inch pulley[31] to the right end of the shaft, file a "flat" on one side, and secure the pulley with the allen screw in the pulley. Finally, run a #62 spring[32] from the former seat hole[8] on the frame to the free end of the grinder shaft assembly.[27]

The adjustable table will be supported by a ¾-inch, 3-foot steel bar[33] with a right angle bend 14 inches from one end. To bend the bar, secure it in a vise and hold a torch 14 inches from one end. Let it get good and hot. When it turns a dull, red color, bend by exercising pressure on the other end by hand— remember to wear gloves!

Cut a 6-inch length of ¾-inch steel tubing[34] to support and guide the table bar[33] you just bent. Weld a ⅜-inch nut[35] to the side of the tube[34] and drill and tap threads through the tubing, guided by the nut, so that a bolt[36] can be screwed through the side of the tubing to secure the table support bar in place. As an option you may want to weld a 2½-inch piece of ¼-inch shaft[37] across the bolt's head to form a "T-bolt" for easy tightening by hand.

Now you're ready to weld the tubing to the frame. Place it far enough forward on the frame to be out of the pedaler's way. The T-bolt should face away from the bicycle, to be readily accessible for quick adjustments. Use a square to align the tubing on a vertical axis. Any misalignment will be amplified by the long table extension. File all burrs from the inside of the tubing to allow the table support bar maximum mobility.

Construct two more bar supports from 3-inch lengths[38] of tubing and weld them together at a 90-degree skew. Slide one onto the table bar support. The other will support the table. Build T-bolts[39] for the bar supports.[38]

We found that a 16 × 11-inch piece of ¾-inch hardwood made a nice table top.[40] Thread a 6-inch length of ¾-inch steel rod 1½[41] inches.

Screw a ¾-inch nut[42] onto the rod, add a washer,[43] and insert the shaft through a ¾-

inch hole in the table. Countersink a thin nut[44] and washer[45] when securing the shaft on the other side of the table. Now you are ready to slip the shaft into the support tubing[38] already on the table support bar.[33]

Scrounge or build a padded seat with a back support. We used two 1-foot lengths of 1¼-inch square metal tubing[46] welded together at a 100-degree angle to support our homemade seat. Two pieces of ¼-inch plywood[47] were cut 12 inches deep and 15 inches wide, padded with foam,[48] and covered with vinyl cloth stapled to the back sides of the plywood. A ⅞-inch shaft[49] was welded vertically to the tubing beneath the seat and inserted into the hole formerly supporting the handlebars.[50] This seat will not be adjustable and should be measured for height and distance from both the pedals and the chuck-table work area to fit your particular needs. If the seat is too close, put an S-shaped bend in the shaft[49] to allow the distance you need. When you are sure it is in a comfortable position, weld the seat in place.

You may find it helpful on some jobs to have toe clips for the pedals. Pick these up at your local bike shop. The flywheel[24] is another option. A lawn mower graveyard is a good place to scrounge one of these. You may not find it helpful for every job, so we suggest bolting[51] it on instead of welding to keep it removable.

Always think safety when using your cycle. The flexibility of the idler should help you avoid pinching your fingers when changing gears. We strongly suggest building guards for the pulleys so your fingers don't get caught in the V-belt.[26] Safe Pedaling!

Rear-wheel Bicycle Adapter
Materials

MAIN FRAME
The following lengths of angle iron:

2   40" lengths[24]
1   5" length[25]
2   4¾" lengths[26]
2   7½" lengths[2]
2   18" lengths[3]
2 self-centering pillow blocks[16]

BICYCLE MOUNTS
The following lengths of angle iron:
2   14" lengths[6]
2   8" lengths[7]
2   7" lengths[8]
4   1½" lengths[9]
2 turnbuckles (one end loop, other hook)[10]

POWER ARM
The following lengths of angle iron:
2   18" lengths
1   5" length
1   4½" length
4 self-centering pillow blocks [13] [15]
1   10" length of steel stock[12]
1   11" length of steel stock[12]
1   6" dia. wheel[20]
4 bushings (optional, if pillow blocks without lodging screws)
2   54-tooth sprockets, #35[17]
2   12-tooth sprockets, #35[17]
length of #35 chain[22]
1 heavy-duty spring[23]

TABLE (see other set of plans)
1   3-foot length of ¾" steel stock[33]
1   16" × 11" × ¾" hardwood board[40]
3   3" lengths of ¾" ID steel tubing[34] [38]
1   6" length of ¾" steel stock threaded 1½" on one end nuts, bolts, washers[41]
3   2½" lengths of ¼" steel stock[37]

Building Instructions

This design adapts any ordinary bicycle to a rear-wheel power takeoff to harness the power-of-the-pedal. For less than $45.00, you should be able to build this simple

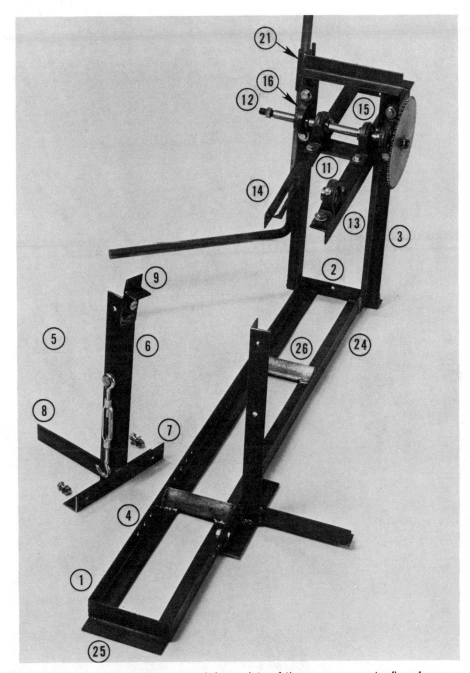

Figure 3-33 The rear-wheel adapter model consists of three components: floor frame and rear uprights, bicycle frame mounts, and spring-loaded power arm.

mounting frame to use your bicycle one minute for grinding grain and the next minute to ride to the store.

Except for a welding or an acetylene outfit, tools needed to build this unit are commonly found around most workshops—drill, hacksaw, file, and allen and common wrenches. Refer to the materials list and photographs regularly for clarity but use this model merely as an example. Let your imagination improve on our design to best fit your specific needs and available materials.

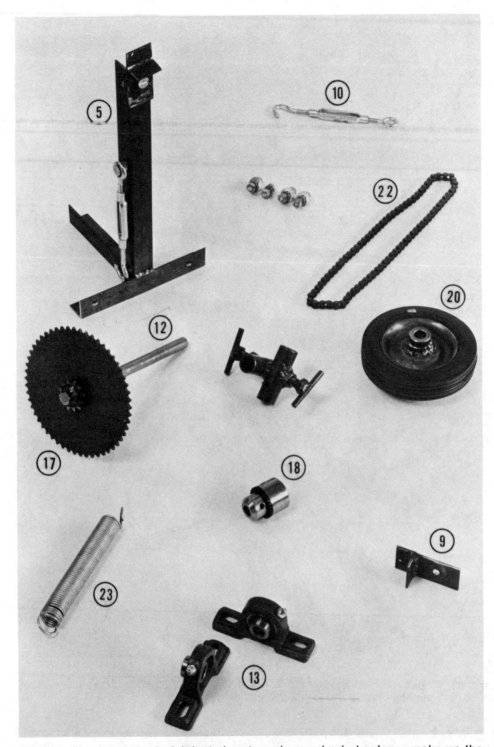

Figure 3-34 A few simple fabricated parts and some basic hardware make up the rear-wheel power adapter.

Figure 3-35 Attached to a five-speed bicycle, this prototype spun the chuck at a rate of over 5,000 rpm's.

Figure 3-36 Bicycle rests on the two frame mounts and is secured in place with two turnbuckles. A converted electric grain mill sits on the adjustable table and its drive shaft is clamped inside the chuck.

Figure 3-37 The Thompsom Mill, manufactured in Mesa, Arizona, can be operated electrically or by pedal. The power takeoff frame is limited in that it will support only a single- or three-speed bicycle.

Construction of the frame is the first step. Using 1¼-inch angle iron, build a floor frame[1] 40 inches long and 5 inches wide. The rear spacing brace[2] should be 7½ inches long to lap outside the frame to support two 18-inch uprights.[3] Before welding, drill ten ¼-inch holes[4] through the sides of the 40-inch[24] pieces at 1-inch intervals starting 7 inches from the front end. *Caution:* Everything must be aligned symmetrically in this model for parallel positioning. Make sure the holes are aligned directly across from their mates. The bike frame mounts[5] will bolt through these holes and can be adjusted to fit different sizes of bicycles.

To build each bicycle frame mount[5] you will need a 14-inch upright,[6] an 8-inch securing bar[7] with ¼-inch holes drilled 1 inch from either end for bolting the mount to the 40-inch floor frame, a 7-inch balancing extension,[8] and a specially constructed frame-rest.[9] Weld the three lengths of angle iron at right angles to each other according to the

photograph. The frame-rest[9] is made from a 1-inch length of angle iron welded to a 1½ × 2-inch steel plate so that a ledge is formed for the bicycle frame to rest upon. A bolt fastens it to the upright. Another bolt fastens a 10-inch turnbuckle[10] to the side of the upright. The turnbuckle[10] should remain loose so you can swing it into place on the bike's frame before tightening.

The power arm[11] is an 18-inch-long, 5-inch-wide frame hinged slightly rear of center on the frame's rear uprights via an axle[12] and pillow blocks.[13] Using 1¼-inch angle iron, space the two 18-inch lengths at the middle and rear with 5-inch lengths. Secure a set of sleeve-bearing pillow blocks[13] at the front end of the power arm in slots[14] rather than holes so the front pillow blocks[13] are left adjustable. Center a second set[15] slightly rear of center on the power arm and a third[16] at the top of the frame's rear uprights.

Next you will need two axles[12] to fit the pillow blocks.[13,15] Thread the end of one

Figure 3-38 The Bik-O-Generator, manufactured by Homestead Industries, will power a grain mill or generate electricity. A pulley attached to a power takeoff wheel could be attached to many other belt-powered tools as well.

with a right-handed thread[12] to fit the Jacob's chuck.[18] Flush with the other end, weld a 12-tooth sprocket and a 54-tooth sprocket[17] separated ⅜ of an inch by washers. Place the axle[12] through the pillow blocks[15] threaded end first and attach the chuck.[18] The other axle[19] should have the remaining sprockets welded in inverse order to those on the first axle. Placing a 6-inch wheel[20] between the pillow blocks,[13] insert the axle[19] through the pillow blocks,[13] wheel,[20] and bushings to secure the wheel. If the pillow blocks [13,15,16] do not have lodging screws, you will need to secure the two axles[12] with one more bushing on the outside of each pillow block.[13,15,16]

See the preceding set of plans for the instructions for building and supporting an adjustable, swiveling table.[27] The supporting tube[21] for this model will be welded high on the frame's rear uprights.[3]

For the finishing touches, adjust a chain[22]

to fit the sprockets. You can loosen the left pillow block[13] and loosen the chain[22] to change gears. Next, drill holes for and attach a heavy-duty spring[23] between the rear end of the floor frame. This will keep the wheel[20] on the power arm firmly wedged against the rear bicycle wheel.

To use your new power adapter, push the power arm[11] to the floor (expanding the spring[23]) and back your bicycle onto the frame. Lift the bicycle so the rear wheel is off the ground and set the frame on the two frame-rests.[5] Swing the turnbuckles[10] around to secure the frame in place and tighten. Let the power arm up[11] and align it against the wheel.[20] Fix your tool in the chuck,[18] secure it to the table[27] or wedge it against the floor. Most jobs will require two people for this setup—one to pedal, the other to operate and feed the implement being powered.

DESIGN

**Out of the drum roar of fancy
out of the rhumb line of chance
comes design.**

James C. McCullagh

CHAPTER FOUR

American Tinkerer: Further Applications of Pedal Power

By John McGeorge

Pedal power was not always a stranger to man. In the years between World Wars I and II, a number of countries developed military field radios powered by cranks which were hand operated. The Italians, who seemed to have a better understanding of the system, built field radios powered by the bicycle generators.

The famous "Gibson Girl," which was used by ditched aircrews in World War II, was a low-powered code transmitter that automatically sent a distress signal on two frequencies when the handle on the top was churned. The unique hourglass shape of the outer case allowed the transmitter to be clasped by the upper legs while the operator applied his muscle to the handle.

Americans in the Far East soon learned that people put pedal power to all sorts of uses, including pumping water, hoisting loads, and cultivating the fields. Occa-sionally, "foreigners," so taken by the prospect of muscle power, lent their services. Legend tells that Geoffrey Pyke, the British inventor who perfected the "Weasel" amphibious tracked vehicle, also invented a pedal-powered tractor for use in China. His tractor, which had seats and pedals for 10 men and was geared down to go slowly, apparently worked very well.

Most operations in the world can be accomplished by less power in a longer time. Power is, by definition, foot pounds of effort produced over a given time. You can get the job done over a longer period with less effort. Calculations show that to travel 10 mph on a 10-speed bicycle will require about eight calories a minute. This is about 120 watts or about one-sixth horsepower.

One-sixth horsepower is the amount of power developed by an electric mixer or a small power tool. It is probably not possible

Figure 4-1 Pedal-driven primemover

to compare the outputs of an electrical motor and the operator of a bicycle. The rating of a motor is generally based on the power delivered at full output and does not actually reflect the consumption of energy required to perform a given task. For example, a washing machine motor might be rated at one-third horsepower even though it normally develops only about one-sixth horsepower while washing clothes. The additional power is designed to take into consideration possible misuse and overloads.

Another factor to be considered is the need to buy a standard frame motor. On the other hand, the human operator can and will supply exactly as much power as he needs to do a job. Pedal power matches the man and the task.

Because of our belief that the human muscles, particularly the thigh muscles, are noticeably underused in American Society, we constructed a pedal-power machine, a primemover. Many of the possible applications of pedal power have been left to the ingenuity of the reader.

Our primary consideration when we designed the machine was to construct a device which would give us a bicycle-powered "motor" to supply energy for various applications. To do this we attempted to provide rotary and reciprocating motions. Long years of experience have indicated that the bicycle form of motions can extract the maximum amount of useable energy from the human body.

The Frame

In order to get maximum power output, the power of the rider must be available at the pedals. This makes it imperative to have the rider properly seated with handlebars to help locate him and to control the torque reaction on his body as he pedals. Figure 4-1 shows the frame with the jackshaft mounted in place of the normal load wheel. Although you could use a standard bicycle frame, our basic frame was an AMF "Whitely" exercise cycle. We removed the bicycle-type wheel, the speedometer, and the friction brake. The handlebars and seat remained as supplied.

The Jackshaft

In a machine shop the jackshaft is commonly used to connect the drive motor to the tools. Old-style mills with a water wheel or one central steam engine had a super-sized shaft that ran the length of the building, the lineshaft. This shaft drove all the various tools in the shop; but with the advent of electrical power, the lineshaft was aban-

Figure 4-2   Close-up of drive assembly and flywheel

doned: the shop with a lineshaft was too difficult to use, the bearings dripped grease, and the belts flapped and were noisy.

However, in home shops the jackshaft is still used. The idea is to make various ratios and output speeds available to the user of the machinery. In our case we had a V bolt pulley on one end of the shaft and our flywheel on the other. The shaft runs on two ball-bearing pillow blocks. The chain sprocket on the shaft has 26 teeth and the pedal crank sprocket has 22 teeth, which means the jackshaft turns at slightly higher speed than the foot pedals.

It was necessary to lengthen the chain as the total circumferential distance is greater. This was done by adding a section of bike chain and another splice link.

The V-belt pulley chain sprockets and the flywheel are hung on the mainshaft, which is 1 inch in diameter and 20 inches long.

To adapt the chain drive to a more nearly 1:1 ratio, pedals to jackshaft, it was necessary to bolt a small bicycle sprocket to a sprocket which was bored out to 1 inch to fit the jackshaft.

The Flywheel

Because of the need for inertia to carry the mechanical operation through high peak torque requirements, we made a flywheel. The idea here is to get the wheel rolling and then use it to carry you through the tough peaks of effort.

The wheel we made for the primemover was a 20-pound chain sprocket with a 1-inch bore and a diameter of about 14 inches. Using a bench grinder, we removed the sharp teeth to prevent accidents. As you see in Figure 4-2, we drilled a number of crankpin holes on the flywheel at various radii to give us a selection of stroke lengths. It is sometimes desirable to be able to lengthen or shorten a reciprocating stroke to match the work required.

The connecting rod end turns on a ⅜-inch

bolt. This bolt can be moved from one radius to another to change the stroke. The bolt is locked into place and the ring end of the connecting rod is put in place. The retaining nut is applied loosely and held in place with a wrench as the wing nut is tightened on top of it to jam it in place.

If the retaining nut is locked down tight on the connecting rod end, it will bind and break.

Making a Flywheel

One of the ways to get more inertia is to make a flywheel, possibly in the range of 25–30 lbs. Get a big cast-iron single-groove sheave. They are available in sizes up to 14 inches in diameter. Obtain a cylindrical container somewhat larger such as a cheese box, hat box, dishpan, or small fiber shipping container.

Plug the shaft hole in the sheave with a dowel or broom handle the diameter of your jackshaft. The dowel should be liberally greased with axle grease or lard. The stick or dowel must be straight and true in the hole. Center the pulley in the exact middle of the container. Mix up 25–30 lbs. of a cement-sand mix and pour it in the container. Be careful to distribute the concrete evenly. Grease the inside of the container to make it easy to get the casting out.

The V-belt Pulley

To get power out for rotary motion applications, we used the V-belt pulley. V belts and V-belt pulleys are available at most hardware stores. By using various pulley sizes on the end of our jackshaft and also on the driven machinery, you can obtain a wide variety of speeds for different applications. It is possible to have speed ranges from a high of about 14 times pedal speed to a low of about one fifth. This would be a maximum of about 840 rpm to a minimum of 12 rpm. This is assuming a pedal rate of 60 per minute, one per second.

Remember also a V belt can make a 90-degree turn in plane. This would enable the primemover to supply power to a device such as a potter's wheel.

Possible Applications
Trash Can Washing Machine

One of the possible applications for a pedal-power primemover is its use as a power source for one of the most time-consuming and unrewarding jobs around the house: washing clothes. The principles in laundry are simple: the dirt is held in the material by grease of some sort. The soap emulsifies the grease and the water washes the dirt away. The trick is to bring the soap in contact with the grease frequently and forcefully. This is almost universally done by shaking or pumping the water and dirty clothes back and forth or up and down.

Grandma's grandma rubbed the laundry with soap and sloshed it up and down in the water. Grandma used a washboard which subjected the clothing to mechanical vibration while it was scrubbed up and down over the ribbed surface.

With the advent of electrical power, the sloshing around was done in a tub. The agitator slopped the clothing back and forth and we had the modern equal to grandma's washtub.

My grandmother had a machine with a cavernous tank. Mounted over the top was a frame with a mechanical marvel which drove four rubber devices that looked like plumber's helpers. These rubber cups pounced up and down at 60 times per minute. The sketch in Figure 4-3 will give you an idea of the bicycle-powered equal to Grandma's washing machine. The rotation of the main flywheel is converted to a reciprocating motion which drives the plumber's helper up and down in the tub.

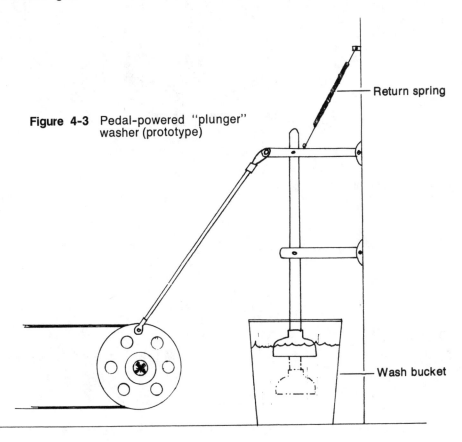

Figure 4-3 Pedal-powered "plunger" washer (prototype)

Return spring

Wash bucket

The parallelogram linkage keeps the plunger aligned with the bucket. This system works well. Five minutes of leisurely pedaling will wash 5 lbs. of clothing. Drain the tub and refill with water. Pedal for two or three minutes more and drain. The clothing will have to be rung out either by hand or by using a roller-type wringer. Sears carries a suitable wringer in the big catalog (11K 5770C).

I remember reading how ranchers, at least those who drive over rough roads in their pickup trucks, wash clothes. They put the dirty duds in a milkcan, fill it three-quarters full with hot water, add soap, tie it out of the way in the front end of the pickup box, and drive. Come supper time the dirt is out and a quick rinse and that's it. The next best thing

is a small trash can (15 gallons) mounted on a set of trunnions. The reciprocating motion of the bicycle primemover is used to rock the can.

With the lid on and locked tight, the water leakage is minimal. The can is equipped with a drain valve, and a garden hose can be attached to run the spent water out to the garden. Detergent and dirt are not harmful to growing plants.

The rocking action causes the laundry to tumble over and over in the water. It is possible to use cold-water detergent and water right out of the pipe. Another trick used in the old days to save water and soap was to wash your white and light-colored clothing in the water first, then your colored stuff, and finally work pants, etc. Liberal amounts of Clorox

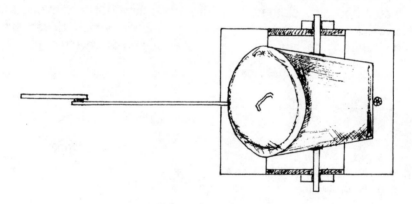

Figure 4-4 Trash can washing machine

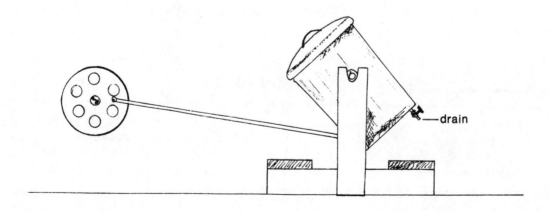

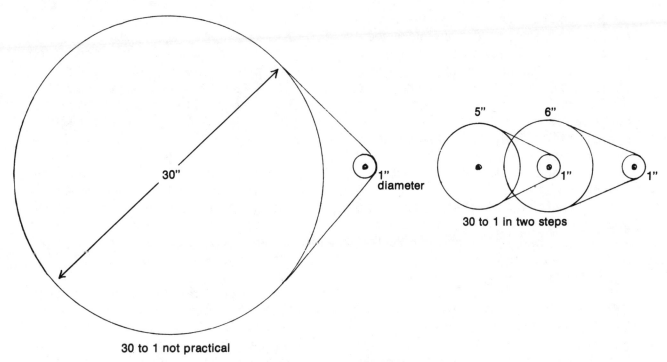

30 to 1 not practical

30 to 1 in two steps

Figure 4-5 Pulley system

were used to keep the cleaning process going. Hot water was in short supply and the Saturday night bath was a cultural institution. Cowboys sometimes anchored their dirty clothes in the bottom of a fast stream overnight.

Driving a Wringer-Type Washer • The non-automatic washer, which was common up till the middle or late 1950s, has possibilities as a pedal-power laundry machine. These wringer machines are available from Sears and other sources. The motor frequently drives a belt system. The pedal-power primemover can be substituted for the motor in this case too.

You must constantly bear in mind whenever you plan to use the primemover as a substitute for an electric motor that you must gear or belt up to a speed which matches the electric motor's normal speed. About 40–60 pedal rotations per minute is a comfortable rate on a bike. The washing machine of this type normally has a 1,800 rpm motor. The pedals' rpm must be geared up by $\frac{1800}{60} = 30{:}1$

Speed Changing • To figure out speed ratios, work like this:

Assume 60 rpm is your normal pedal speed. Most people will find that 50–60 rpm is comfortable. If you can find the machine's motor name plate, it will give you the horsepower rating and rpm. Don't attempt to power a machine which has more than a one-third horsepower motor. No one can produce that much power except in Olympic effort. Stick to jobs of one-fourth horsepower or less.

If the rpm of the motor is more than 1,800, the gear up ratios will be too great to operate. As the ratio increases, so do the losses and it can reach the point at which the losses in the belts, pulleys, and bearings are equal to one-half the power developed. Look over the mechanism which you intend to drive and see if you can get into the drive train someplace that gives you a direct ratio.

93

For example, I have a Polish-made grain grinder which is designed to operate at about 60 rpm. The motor I had was a 1,800 rpm so I had to step the motor down by 30:1 to go to 60 rpm. Now, to get 30:1 you basically need a 1-inch pulley. Generally speaking, this is out of the question because a 30-inch pulley is large and a 1-inch pulley is so small the belt has too small a radius to curve around and not enough friction area to effectively transmit power.

The answer is a two-step reduction, 1:6 and 1:5, which is easy to do. The losses here are greater than with a single reduction; however, the system is much more compact.

To replace the motor with a pedal system here would be ideal. A direct one-to-one drive will do nicely. It is not wise to go up to motor speed with a speed increase and then back down again by a like amount when you should have gone directly or nearly directly. Analyze your problem.

The Wood Saw

We used the simple reciprocating motion of the main wheel to power a wood saw. Figure 4-6 shows a prototype for the system.

The wood is held in a V-shaped cradle; the cradle and saw must be rigidly linked or securely staked to the ground. The main saw arm is guided to prevent the saw from leaping out of the stroke. We chose to use the common Swedish-type bow saw because it is designed to operate at low surface speeds. A circular-type saw might work in light work application, but the feed rates would have to be low or you would bog down the rider.

It is possible to modify a chainsaw chain to be driven by the primemover. However, because the saw is designed to operate at faster cuts than the handsaw, it might prove more difficult to cut the same amount of wood.

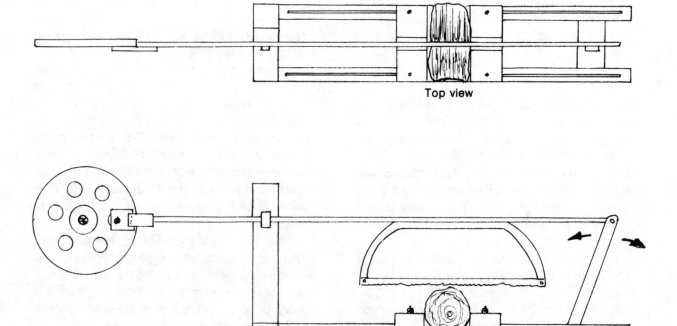

Top view

Side view

Figure 4-6 Pedal-driven wood saw (prototype)

Water Pumping

Water pumping is a basic energy user on the farm. As was shown in *Organic Gardening and Farming*, it is possible to pump a considerable amount of water with pedal power.

Many house water systems use shallow well pumps. Note that this pump has a V-belt pulley which makes it possible to connect to the primemover. This type of piston pump can be operated at lower speeds and powers. The system might require five minutes of effort four or five times per day to supply minimum household water needs. The system pressure and the depth of the well would determine the effort required.

The Pitcher Pump

Many farms or rural locations have perfectly usable pitcher hand pumps. The pump will pump water with little effort if it is properly set up. An actuator rod from the reciprocating output of the primemover can supply the up-and-down motion to pump water, as shown in Figure 4-7. Because the

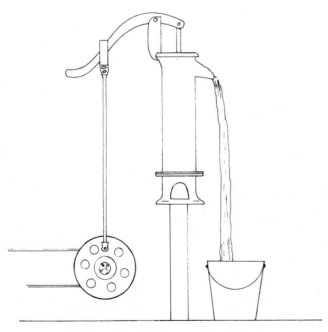

Figure 4-7 Pitcher pump (prototype)

legs are stronger than the arms, it will be possible to pump more water in considerably less time with less effort than before. Force-type pumps, which resemble a pitcher pump, are available too. A force pump not only lifts water from the well but also supplies it at pressure so it may be pumped to a storage tank in a home, barn attic, or loft. A force-pump system might be useful for washing animals or farm machinery.

The primemover could also operate an old refrigerator compressor which can be used to supply 40–50 lbs. of compressed air to inflate tires. Many old refrigerators have a separate compressor unit which can be used as a high pressure air pump.

Log Splitter

Hydraulic log splitters are available in kit form. In Figure 4-8 note that the various components may be purchased separately. The hydraulic pump is designed to be driven by a 3- to 5-horsepower gasoline engine. By using pedal power it is possible to generate the needed horsepower. The maximum power is required at the first push. If the primemover had a selectable series of ratios as a 10-speed bicycle has, it would be possible to make best use of the operator's efforts.

A simpler system can be devised using an automotive-type scissors jack. The jack, which can exert 4–6 tons, is driven by a tumbler shaft from a reduction jackshaft belted to the primemover (see Figure 4-9).

Cider Press

Another application related to the log splitter is a cider press. The chopped apples are bagged in muslin sacks about 12 inches square and sandwiched between 12 × 12 pieces of board. The press must have a tray below to catch the cider. The pomace left over can be used to make pectin or as animal feed. The apple chopper could be pedal powered also. Applications requiring use of the strength of the arms can usually be powered by pedal power.

Applications of pedal power are as extensive as our imaginations permit. Hopefully, some of the preceding suggestions will encourage the reader to extend the horizon of pedal power. All of us, in both developing and developed countries, will surely benefit.

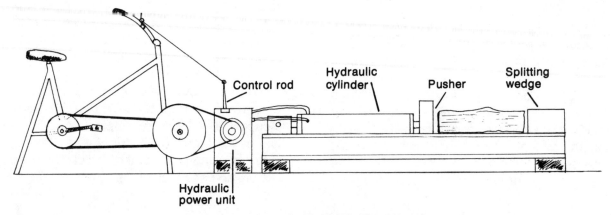

Figure 4-8 Hydraulic log splitter (prototype)

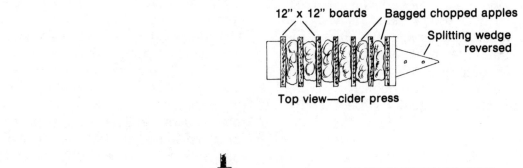

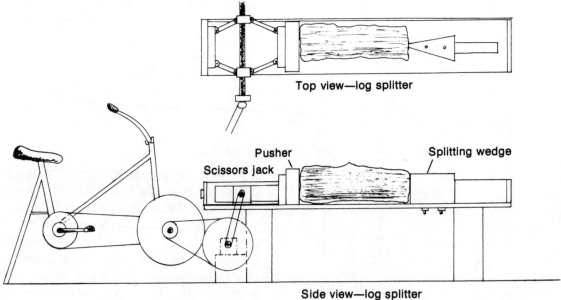

Figure 4-9 Scissors jack log splitter and cider press

MAN

Between the plow and the combine
the bicycle and the truck
the forest and the rock
man moves in his sweat
like the sea.

James C. McCullagh

CHAPTER FIVE
Treadle Power in the Workshop

By Mark Blossom

With motor-driven power tools, one can work quickly and with little physical exertion. Many craftspeople feel that without them they could not compete in the marketplace. Despite their advantages, however, I prefer not to use power tools. I find their noise and insistent high speeds unsettling and antithetical to the tranquil frame of mind I seek while working with wood. Although I am not yet skillful or patient enough to dispense with power tools altogether, it is my goal to have a completely "people-powered" woodworking shop.

My first step in this direction was to convert an old band saw to treadle power. In my design, the treadle turns a massive flywheel, which, in turn, drives the band saw. I operate the machine with one foot on the floor and the other pumping the treadle. As wide as the treadle is, I can stand directly in front of the blade or off to either side and still be able to pump. This feature is useful when cutting a long or large workpiece. A wide treadle is in this case more versatile than bicycle-type pedals, which confine the operator to one spot. By using both legs, as with pedals, however, one could probably generate more power than with only one leg. In my design, power is applied to the treadle only on the downstroke. The heavy counterweighted flywheel evens out this periodic force and keeps the saw running at a steady pace.

Hand-Made Toys

Band saws are designed for making curved or scroll cuts. I have used mine extensively to make wooden toys from half-inch and thinner stock. This kind of light work is very easy and pleasant. I also routinely step up to my band saw to make straight cuts in boards that one would ordinarily cut with a handsaw. I can definitely cut more quickly

Figure 5-1 Treadle-driven machines allow for delicate workmanship.

and with less effort using leg instead of arm muscles.

I appreciate having direct control over blade velocity. I can speed up for more power when cutting tough material and slow down for intricate designs that require concentration. This is an advantage over motor-driven band saws that you have to stop and adjust the pulley ratio on in order to change speeds.

My machine is capable of handling some very thick wood. The primary limiting factor is the strength of the operator. I've made chair rockers out of 1½-inch black walnut and brackets out of 4 × 4 western red cedar, but these jobs leave me exhausted. The flywheel is definitely a help when cutting heavy stock.

With a band saw one can turn abundant native materials into saleable items. Small tree branches can be sliced into interesting buttons. They are finished by drilling two holes and sanding. Black walnut slices likewise are popular with handicrafters, so popular in fact that plastic imitations are on the market! As you slice up walnuts, you will also be accumulating a pile of sliced

Figure 5-2 "Toy Man with a Hoe"

Figure 5-3 "Toy Man with a Hoe" and background

Figure 5-4 "Dancing Ladies"

Figure 5-5 Walnuts, buttons, end pieces, and nutmeats fashioned on the band saw

Figure 5-6 Full view of band saw

nutmeats, just the right size for cake or cookie recipes.

The base of this machine is 3 feet square, considerably larger than a comparable motor-driven tool. It needs to be this large to accommodate the flywheel and to provide stability despite the reciprocal action of the treadle.

I believe that there is more of the essence of a craftsperson in his work if the energy that went into it comes from his own body. I identify more with things I have made with my own energy than with things made with the help of electricity. Operating a band saw can be hard work, but it is a good way to exercise while working indoors and a productive way of keeping warm in the winter.

Construction

This design was worked out with an eye toward using the materials I already had on hand. Everything, except for some of the bolts, was second hand or salvage. Substitutions from the reader's own scrap piles will suggest themselves. Probably not many readers will have an old band saw to start with, but the treadle-power base could be adapted to any other tool with a similarly positioned pulley.

There are a number of companies that sell kits and parts for making your own pulley-driven tools. Two such companies are Gilliom Manufacturing Co., 1109 N. 2nd Street, St. Charles, Missouri 63301, and American Machine and Tool, Royersford, Pennsylvania 19468. Others can be found in *Popular Mechanics* and *Popular Science.* Although I have had no personal experience with their products, they do seem to offer the do-it-yourselfer a way to put together his or her own tools at a savings. In some cases

they offer two versions of a kit; if the more expensive one has better bearings, by all means choose it. One thing I learned early is the importance of good, low-friction bearings. Whatever friction there is in the machine will detract from the amount of your energy that is actually delivered to the workpiece.

The flywheel in my machine is from an old (late 1950s) Bendix front-loading home washing machine. The part that I used was a lens-shaped object about 2 feet in diameter that was actually stationary in its original application. The front of it served as the back wall of the washing chamber. It has a shaft,

Figure 5-7 Front view of flywheel and frame

Figure 5-8 Side view of flywheel

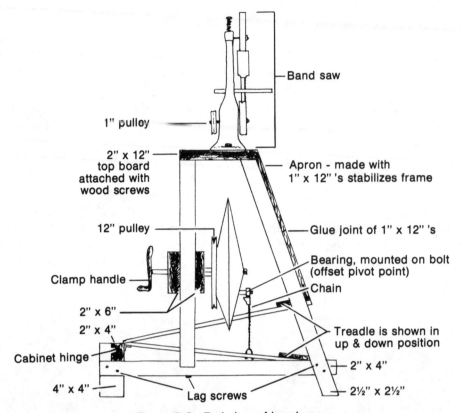

Figure 5-9 End view of band saw

mounted on bearings, through the center of it. This shaft had the perforated washing basket on its front flange. The rear, threaded end of the shaft now passes through the wooden crosspiece of the treadle base and is anchored firmly in place by the nut that originally held the pulley. Thus, the shaft is now stationary with the flywheel rotating on it; whereas originally the flywheel was a stationary base with the shaft rotating in it. I think it is remarkable how a junked washer, relic of an opulent age, yielded up parts for a new tool in the post-industrial renaissance!

The flywheel was originally hollow, the two walls being stamped sheet metal. In order to make the offset pivot point, I drilled a hole through both walls, 4 inches away from the shaft center, and inserted a bolt with about 2 inches of thread protruding on the front side of the flywheel. Some flat washers and a nut

hold the bolt in place and give clearance to the ball bearing that fits on the bolt thread and is held in place with a final nut. (If you have a ball bearing with an inside diameter slightly larger than a standard bolt size, the difference may be made up by wrapping the bolt tightly with paper or leather.)

I wrapped the outside of the bearing with a strip of thick leather and then tightened a hose clamp around that. Next I passed a heavy wire between the outside of the bearing and the hose clamp at the point where the ends of the leather strip meet. This wire connects with the chain that comes up from the treadle.

A 12-inch pulley is attached directly to the back side of the flywheel. The pulley is open in the center so the flywheel shaft can pass through unimpeded. Suitable pulleys can be found on junked ventilation and laundry

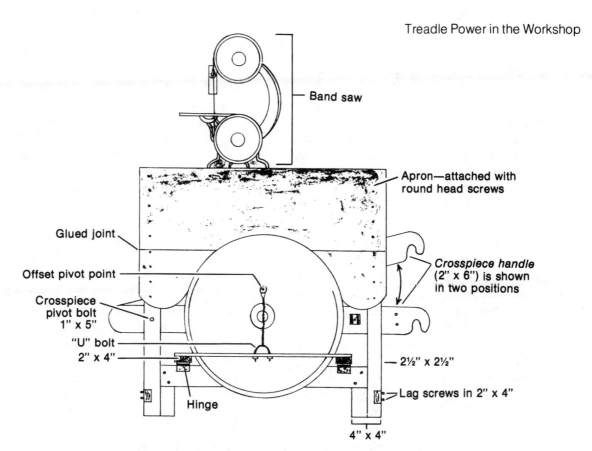

Figure 5-10   Side view of band saw

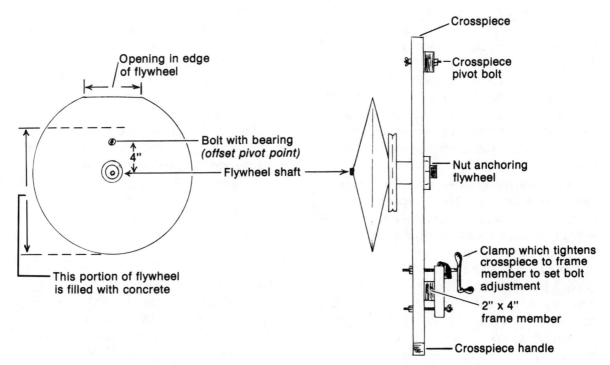

Figure 5-11   Left—flywheel; right—top view of flywheel and crosspiece

equipment. The 12-inch pulley could also be made of wood or by attaching two same-diameter sheet metal discs together by a ring of rivets about an inch in from their edge and then flaring their edges apart to accommodate the V-belt. This pulley could also be larger or smaller, within the range of 10 inches to 14 inches; of course, this will have a proportional effect on the operating speed of the tool.

I next cut an opening in the rim of the flywheel. This opening is placed so that the offset pivot point is on a line between the shaft and the opening. With the flywheel standing on edge and the opening uppermost, I dropped in pieces of lead obtained by breaking up two old car batteries. I poured concrete in after it, until the flywheel was about three-fourths full.

With the flywheel thus counterweighted, the down stroke (power stroke) on the treadle lifts the counterweight up and around. It continues on past the apex by its momentum and drops back down, lifting the treadle for another down stroke. The counterweighted half of the flywheel is about 30 lbs. heavier than the half with the offset pivot point. The entire wheel weighs about 100 lbs.

I don't expect many readers will be able to obtain an appropriate Bendix frontloader, although there must be thousands of them in junkyards and basements across the land. It may be that other brands and models would serve.

There are many other combinations that would work as well. The flywheel could be formed of reinforced concrete, with blocks of wood imbedded for attaching the central and offset bearings. It could be of hardwood, counterweighted with steel plate semicircles. Other possibilities include an industrial flywheel from the scrap metal yard, an old grindstone, a manhole cover, or the flywheel off a large engine which might already have a shaft and bearings. A flywheel that is thicker in the middle than at the edges stores energy better than one of uniform thickness.

In my design, the flywheel shaft is anchored on only one end. One might be able to use parts from an automobile front wheel assembly as the basis for the flywheel, since they are also attached on just one side.

It would be possible to have the flywheel rigidly attached to its shaft and have bearings on both ends of the shaft. In this case, the shaft would have to have an eccentric or crank formed into it between the flywheel and the front bearing, as well as a connecting rod down to the treadle. If you use this approach, you will have to devise a way of changing the distance between the two pulleys so as to be able to change belts and adjust belt tension.

In my design, the crosspiece that the flywheel shaft is anchored to is pivoted to one frame member and clamped onto another. By loosening the clamp and raising or lowering the free end of the crosspiece, the belt can be adjusted or replaced.

Driven by the 12-inch pulley, the belt passes through a slot in the top board of the frame, to which the band saw is bolted, and drives the band saw's 1-inch pulley. The operating speed of the band saw is thus 12 times that of the flywheel. At 60 strokes per minute, which is a comfortable treading speed, the band saw runs at 720 rpm. This speed can be reduced by replacing the 1-inch pulley with a larger one.

The treadle is attached to the frame with cabinet hinges. Most of the joints of the frame are notched and all are held together with screws or bolts, so that they can be tightened or the frame taken apart and modified if necessary.

Renaissance of Hand Crafts

When foot-powered machines were put to rest at the turn of the century, most people thought that they would never be missed. But now, after many generations during which tools, skills, and old methods have almost entirely disappeared, craftsmen are reconsidering foot-powered lathes and saws. The American Village Institute in Selah, Washington, has constructed a line of foot-powered machines with the craftsman in mind.

Pedal power holds great promise for application in useful work and transportation, but it also has significance in the home workshop for those who desire complete control over their craft.

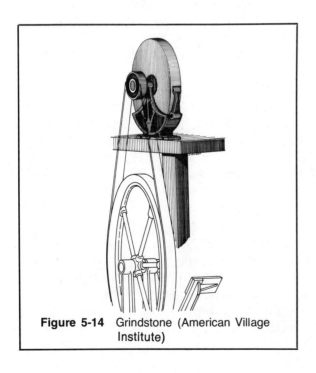

Figure 5-12   Wood lathe (American Village Institute)

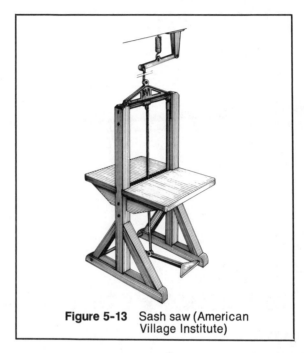

Figure 5-13   Sash saw (American Village Institute)

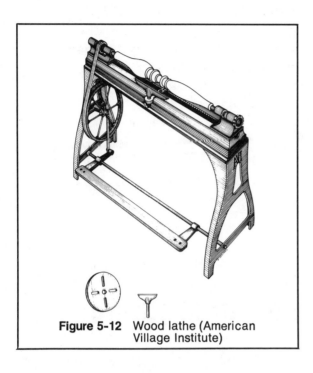

Figure 5-14   Grindstone (American Village Institute)

Figure 5-15   All-purpose foot treadle (American Village Institute)

CONTEMPLATING LEONARDO'S BICYCLE

In the animal of your sketch
pen lines are tubes to the sprocket
spun by the flywheel weight of man.
The T-bar of your fancy
is a rudder of thought
which thrills direction
with a contemplated plane
where little heat
blackens the roadway.

James C. McCullagh

CHAPTER SIX
The Future Potential for Muscle Power

By David Gordon Wilson

Western industrial man—and woman—has surrounded himself with, and perhaps surrendered himself to, a variety of powered gadgets beyond the wildest dreams, or even the desires, of our grandfathers. The power levels of some of these new necessities of life would be staggering even to those of our forebears who harnessed the forces of nature and who planned on a broad canvas—Leonardo da Vinci, for instance. Would even his imagination conceive that a single individual would drive a chariot equipped with the power of over 200 horses—and that this individual would do this for the most capricious of purposes, for journeys of down to a hundred yards?

While Leonardo would without doubt be fascinated with the beautiful mechanical designs which we have developed, even his unfettered brain would have difficulty comprehending why we act the way we do.

He would find, for instance, that some of the people who seem incapable of walking the shortest distance if they can possibly go by automobile try to compensate for the lack of exercise that their way of life entails by pedaling on a device especially designed to incorporate large frictional losses. After a "workout" of a few minutes in the morning on such an exercise "bicycle," modern man is likely to shave with a noisy razor powered by an electric motor, to polish his shoes with an electric buffer, and to use the apartment elevator rather than the stairs even if he has to descend only one floor.

It would be easy to recite, *ad nauseam*, examples of how we have come to depend on mechanical power in various devices in all facets of our daily lives. I believe that it is more helpful to show that these uses of mechanical power fall into three well defined classes and that an intelligent application of muscle power could reverse historical trends and roll back the use of external power in at least one of these classes.

High-Power Devices

High-power devices include airplanes, buses, trucks, tractors, supertankers, subways, sewage pumps, and so forth. We depend on mechanical power for these and many other consequences of our industrialization. One may question the desirability of a way of life that has made these things into necessities. But given our present way of life, there seems little alternative at present to the continued use of these devices in much their present form.

Low-Power
Special-Feature Devices

Then there is a category of uses of power in which the power levels are low enough to be supplied by human muscles but where the use of an independent energy source confers some particular advantages. This advantage is usually that greater control can be given. For instance, an electric pistol-grip drill delivers only a few tens of watts to the drill bit, but its use permits both hands to be used for control and guidance and for applying pressure to the bit. It can make holes closer to a wall or to a corner than can a manually powered drill. The drill bits are less likely to break without the side forces which handcranking imposes. One can control an electric dry shaver much more closely than one can a shaver which has to be squeezed continually to maintain the momentum of a flywheel. The electric typewriter I use enables me to type (and to make errors) at twice the speed at which I could type on my manual machine.

In this class, therefore, I am grouping those low-power devices on which the independent energy source confers a significant advantage over the manual alternatives presently available. It may be, of course, that we could devise better methods of applying muscle power to these devices—of which more later.

Low-Power "Convenience"
or "Status" Devices

The third class is of most interest because it contains those applications where independent power involves marginal, or sometimes negative, benefits. Some automobiles, for instance, have motors which rotate the headlamps out of sight for cosmetic reasons. External damage, ice, or system failure can prevent the headlamps being available when needed, so that on the whole the power feature confers a safety penalty. The automobile is becoming filled with powered windows, powered seat adjustments, powered antenna winders, and so forth, which have some small benefits when they work well (and some safety problems); when they fail, as they are bound to do, they become sources of great annoyance.

I have many other candidates for this class, but I know that not everyone will agree with me. In many cases, the classification choice between the "useful" or the "cosmetic or convenience" categories will depend on the scale of use. Electric can openers, pencil sharpeners, and erasers I would normally place firmly in the "cosmetic" class. But I suppose that if I worked in a restaurant kitchen or in a school and had to open 20 cans or sharpen 20 pencils at a time I might appreciate the powered devices.

It is the same way with lawn mowers and snow blowers. We have all seen someone lugging and straining to get a heavy power mower on to a pocket handkerchief of a lawn, expending more effort than he/she would by pushing a hand mower. Similarly for snow blowers. However, if one has to mow a half-acre of grass, or clear a 50-yard driveway of snow, a power mower and a powered snow blower are each undoubtedly more desirable than their presently available manual alternatives. Lawn sweepers, mopeds, tire air pumps, and sewing machines are also, I believe, in this general class.

Need for Improved Muscle-Power Delivery Systems

I emphasized earlier that sometimes powered devices were more attractive than their presently available manually powered alternatives. In part this is due to a well known phenomenon. Manually powered devices were developed first and reached a tolerable level of convenience. They then tended to stagnate. When designers, usually from outside the industry concerned, applied independent mechanical or electric power, they did so with all the benefits of modern materials and styling. They hired market research people to determine public reactions. The response of the traditional manual-devices manufacturers has generally been to stick stubbornly to the old designs, or to give up entirely. Bicycle manufacturers are a prime example. When bicyclists began switching to motorcycles and automobiles at the turn of the century, development of bicycles virtually ceased. Every year, new car models are announced with new features which in the sum have amounted to enormous improvements over cars of 75 years ago. In contrast, there has been no fundamental improvement in bicycles in this whole period.

To a lesser extent, the same is true with regard to some of the powered systems in automobiles themselves. Manual windows have awkward rotary knobs and cranks, always seeming to be in the way of one's knees and always sticking or slipping; yet they have been used for decades. Manual seat adjustments seem always difficult to locate, awkward to operate, and again liable to sticking. In contrast, modern, well engineered and styled powered systems are obviously attractive, at least to some people. If the same effort had been put into redesigning and styling the manual systems, they would have had wider use at present than they do.

With these general considerations in mind, I want to encourage you to let your imagination dwell on possible future developments of muscle-power applications. I will start with three areas in which I am personally involved, and progress to some blue-sky dreams of the future.

A Recumbent Bicycle

For many reasons, I became interested in bicycles on which one half-sits, half-reclines. One reason was that after a succession of bicycle accidents I had taken up sculling (and making rowing shells) as an alternative way of getting exercise, and I was intrigued with the comfort and grace of the sliding-seat rowing motion. At about that time a Dr. J. Y. Harrison in New South Wales published the results of some research which showed that by modifying the rowing motion, subjects could give out about one-eighth more power than they could in bicycle pedaling and could maintain this increment over the whole period of the tests. Then my bicycle accidents, and those of many others, had convinced me of the dangerous exposure that results from the standard hunched-over, head-forward position. I organized a competition for improved designs in man-powered land transport and the judges selected recumbent bicycles as winners. I later became aware of earlier developments—of the French "Velocar," a recumbent bicycle introduced in 1932 and banned by the International Cycling Union after it broke all track records; of a Scottish subsequent design; and of one by Dan Henry of Flushing, New York. Despite the enthusiasm of the few proponents, the prevailing orthodox view was that the traditional "racing-bicycle" position had been reached after years of trial-and-error in which it had proven itself superior to all others.

I would have probably done no more about it had not a Fred Willkie written from Berkeley to ask me to suggest an interesting and novel design of bicycle, as he wanted to build one. I sketched a recumbent and contributed a small amount of money for materials from a fund given me by Dr. Paul Dudley White for research on muscle power, in return for the promise of a report. To my pleasure and surprise, Fred Willkie made the bicycle and tested it—and he didn't like it. He found the near-horizontal position of his legs and the upright attitude of his back combined to cause his knees to hurt and his hamstring tendons to become sore. He asked for other suggestions. I drew out another recumbent, with the legs lower, the back able to lean

Figure 6-1 Fred Willkie on his "Green Planet Special I"

back farther, the front wheel under the knees, and the handlebar beneath the seat.

This turned out to be a decided improvement, but it was still somewhat painful to ride. I purchased it from Fred Willkie and began making a series of small changes—shortening the wheelbase, making a fully supporting woven seat, and rebuilding the wheels to use larger-section tires. I have been delighted to find that it soon became not only more comfortable than any other bicycle I have owned, but that it had other advantages, foreseen and unforeseen, in addition. Here are some:

(1) One can push directly from the hips or the shoulders; the arms and back can stay relaxed even during maximum effort.

(2) The diaphragm is free to rise and fall (in the crouched-forward position it isn't) and one can breathe more deeply and more slowly at the same level of exertion. For probably the same reason, one's digestion seems better on long rides than it is on standard "10-speed" bicycles.

(3) One can communicate one's intentions to other road users much more easily from the sitting position.

(4) Sitting when waiting at traffic lights is vastly more comfortable and relaxing than being perched on a saddle trying to maintain balance with one pointed toe on the ground.

(5) One can keep pedaling around corners with no fear of a pedal catching the ground, or even a curb.

(6) Having one's legs out in front makes one feel better about a frontal collision. Sitting on a full-support framed seat (with a roll-bar) makes the prospect of being run into from the rear also a little more palatable.

(7) Skidding or otherwise falling off is safer. One generally rolls gently onto one's elbow, hip, and shoulder and one can keep the head out of harm's way. It's quite different from the sudden impact of the whole body or, worse, of the head alone, which can occur from the standard racing-bicycle position.

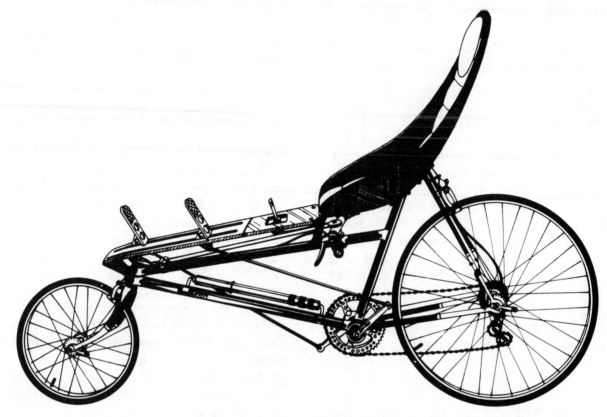

Figure 6-2 Wilson reclining bicycle

(8) With improved brakes (which we have) it is possible to obtain emergency braking at almost automobile levels. In a standard bicycle, one goes over the handlebars if one tries to decelerate at much over a half "g."

(9) In a catastrophic failure of either wheel, or the fork, one is likely merely to be deposited on the ground in the seat frame, which can act as a skid. Similar failures in standard bicycles can be fatal.

The recumbent has some additional minor advantages, and one or two comparatively minor disadvantages. One must carry a flag or other high-up indicator so that one can be seen (there is an improved field of view to see, but decreased reverse visibility). It is difficult to wear effective rain-proof clothing. It seems necessary to incorporate some shielding on the bicycle. This would bring improvements in collision protection, in lower wind resistance, and in increased visibility to others, at a small cost in weight.

I have dwelt on this bicycle at some length because it illustrates the difficulty of coming to any conclusions on new or different applications of muscle power. Having experienced this recumbent bicycle, I have no doubt that it represents not only a current improvement over standard bicycles, but a prospect of further and continual advances. It was denied earlier introduction to general users partly because of a highly conservative, unenterprising industry and partly because of blinkered antagonism among the bicycle-racing community. I believe that it could open up a new group of people to the benefits of bicycling, people who find existing bicycles rather forbidding and uncomfortable.

A Pedaled Lawn Mower

At about the time that Fred Willkie was experimenting with my recumbent designs, Michael Shakespear chose to do his undergraduate MIT mechanical-engineering thesis on a topic which had intrigued me for a long while: the design of a lawn mower which would be pedaled rather than pushed. He turned out to be a superb designer and craftsman, and he produced a machine which was good looking and practical. It had a three-speed transmission and a differential drive to the wheels with a transmission brake. By squeezing a handlebar lever and then pulling back on the handlebars, the whole cutting assembly could be lifted off the ground when traversing paths and the like.

The mower suffered from two drawbacks. One was the great weight of the old cast-iron second-hand components which we had of necessity to buy. The second was the feet-high seating position, which was due to my influence. We had not yet learned from Fred Willkie's experience with his first recumbent that a few degrees in leg angle seem to make a large difference in pedaling comfort and power output. I believe that a new design using the pedaling position we arrived at in the second recumbent, and lightweight construction, would have great attractiveness for people with small-to-medium-sized lawns, a desire for exercise, and a dislike of the noise and fumes of gasoline-powered mowers. Rodale Press is currently funding another MIT student, Lee Laiterman, to undertake just this type of development. His preliminary drawings are represented in Figure 6-3.

Based on tentative calculations, the following specifications will apply: wheelbase—45

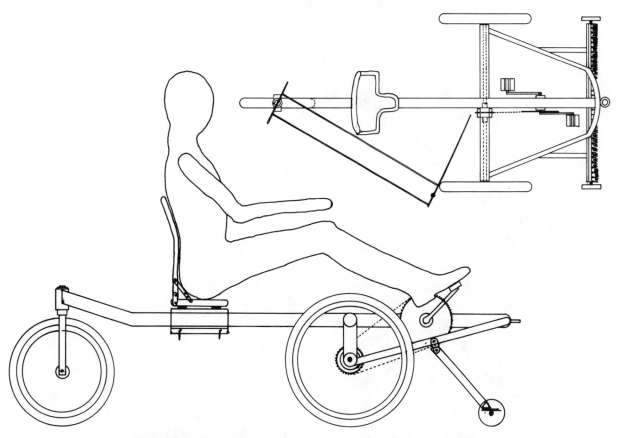

Figure 6-3 Preliminary drawing for pedal-powered lawn mower

inches; front track—35 inches; seat height—19 inches above ground; wheels—2 to 20 inches diameter in front, 1 to 16 inches diameter in rear; cutter length—39 inches; front wheel drive—differential unit mounted on front axle; steering—rear wheel; overall length—7 feet.

The seat mount will be able to slide along the body, thus enabling adjustment for different sizes of riders. The cutter assembly is experimental. If it works, it will cut like an electric hedge trimmer.

An eyebolt is included in the front end to facilitate storage on a wall. The seat back has an adjustment tilt for comfort. It can swing all the way down to facilitate wall storage. The seat is constructed of nylon webbing over a tubing skeleton. The overall weight of the lawn mower will be approximately 45 lbs.

A question to be answered in this work is to what extent the loss of energy which must necessarily accompany the movement of a wheeled vehicle over relatively soft ground can be reduced by good design. It is obviously better to have a slow vehicle cutting a wide swath, rather than the reverse. Or the vehicle could be moved only occasionally, and the cutting mechanism could be powered and controlled as a satellite. The same question and the same range of possibilities arise in snow clearing and in tilling the

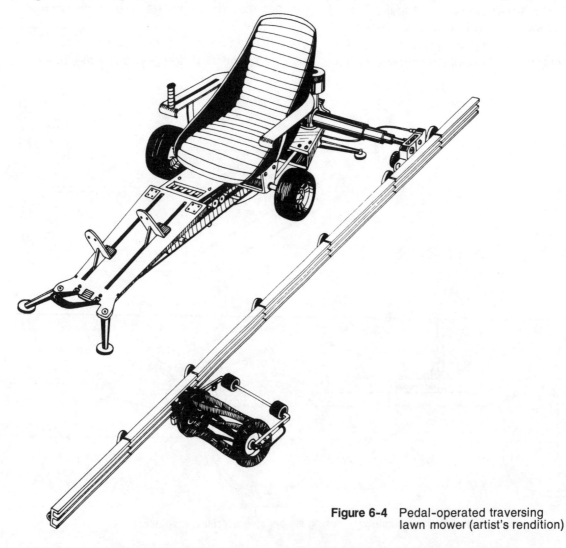

Figure 6-4 Pedal-operated traversing lawn mower (artist's rendition)

ground. The power unit can be stationary, as was the case when traction engines were used for plowing—the plow being dragged across the field by cables running around pulleys which could be advanced by increments. The arrangement has been revived by modern experimenters with computer-controlled plowing systems and by Rodale Press in experiments with its Energy Cycle unit. It will take very skillful design to produce a system which is as acceptable as the self-contained vehicular mower, but the gains in energy efficiency are potentially so considerable that the attempt must be made.

We see the same phenomenon in mowing, tilling, and snow clearing identified earlier as a cause of the demise of so many manual operations. The design of the manually operated tools remained unchanged in their centuries-old forms while continual improvements were made to the new power-operated versions. The same is true with regard to small-boat propulsion.

Pedaled Boats

Oars can be used to propel boats with grace and considerable speed on smooth rivers and lakes. In choppy waters, oars are inefficient and even dangerous. They must be dug into the water at a sharp angle to ensure making connection, because missing—or "catching a crab"—could capsize a rocking boat. So the oars and the required stroke must be short, and the legs can no longer be used as the principal actuators. Instead, the arms and back do the work, with the legs being merely struts.

However, even the most efficient of rowed boats, the lightweight sliding-seat racing shells, have been beaten handsomely by screw-propelled pedaled boats. Oars can be about as efficient as screws while they are in the water at right angles to the boat. But they spend much of the stroke at an angle to the boat, wasting energy. Then their kinetic energy must be dissipated by the rower's

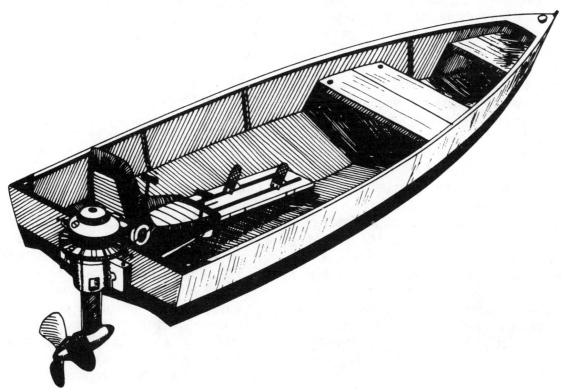

Figure 6-5 Pedal–operated boat (artist's rendition)

muscles, the oars must be brought back fast against the wind, the rearward kinetic energy must again be dissipated by muscles, and only then can another power stroke begin. It is small wonder that pedaled boats can be more efficient. Why haven't they been developed? All the energy of designers and of advertisers has gone into creating a demand for internal-combustion-engine or battery-electric power units, even for applications where a pedaled device would seem to be superior. For instance, to carry a lead-acid battery together with the motor and screw to a boat for a day of troll fishing is hard work and unpleasant. The speed which can be given is small, and there is always the danger that the battery will give out when one is far from shore. Moreover, troll fishermen can go out on days when the winds and low temperatures of open boats can be dangerous, and when the possibility of exercise would be very valuable.

Another example is dinghy propulsion. Yacht owners usually moor in fairly open water and ferry themselves, their passengers, and supplies to and from the yacht in small dinghies. When the wind is high and the water is choppy, rowing can be difficult and a little dangerous, and most yacht owners have changed to carrying small outboard motors. These require the hauling around of a can of gasoline, and they involve the danger of being put out of action by infiltrating water. The potential light weight and reliability of a pedaled "outboard" power unit have an appeal.

Bob Emerson took on the design of such a unit for his bachelor's thesis at MIT. He did not manage to perfect it, but he made an excellent start. Figure 6-5 shows how the entire propeller unit could be turned for steering. The drive is usually the popular system of cables running around sprag one-way clutches. This does not give optimum energy transfer but has advantages in ease of starting and general flexibility.

Yacht Battery-Charging Generator

Dr. Stephen Loutrel likes to sail each summer from the cold waters of New England to the colder waters of northern Labrador, or a similar distant coast. The yacht batteries must be kept charged for lighting and radio and navigational instruments. Being a true yachtsman, he does not like to start his engine except for emergencies, nor to use precious fuel just to charge batteries. Moreover, he goes to these northern coasts for peace and solitude, and shattering the silence with the roar of an engine offends him. He and his wife Liz and their student crew need exercise, which they don't get cooped up in the confined cabin and deck of the small boat, so he has set his students the task of designing a pedaled generator. The first model is rather similar to the Rodale Energy Cycle unit. This configuration is not ideal for use in the tiny cockpit. A better design would use a rowing-type motion, which could, in good weather, use the small foredeck. It would, being almost flat, occupy less space when folded and stored than does the high framework of the existing model.

Irrigation Pumps

Present irrigation devices used in developing countries and powered by human muscle power generally use the arms and back in swinging motions. There are a few examples of the leg muscles being used, particularly in Asia (see chapter two). Some different modern designs for foot-powered pumps suitable for local manufacture are an Archimedian screw driven from a bicycle frame with standard pedals; an endless-chain high-lift pump designed by VITA (Volunteers for International Technical Assistance), with the chain going over the untired rear wheel; and a rocking-pedal diaphragm pump designed by the Rice Research Institute in the Philippines.

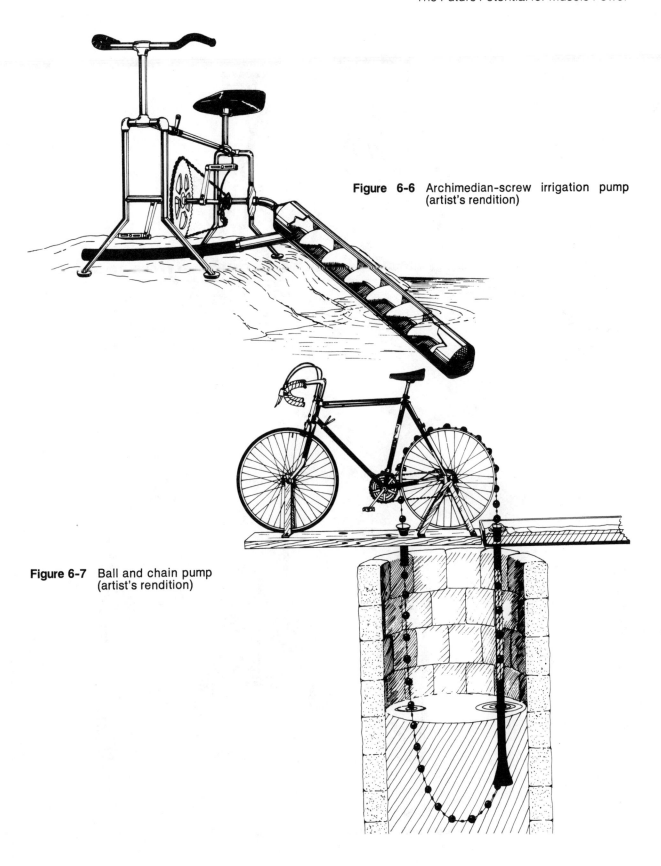

Figure 6-6 Archimedian-screw irrigation pump (artist's rendition)

Figure 6-7 Ball and chain pump (artist's rendition)

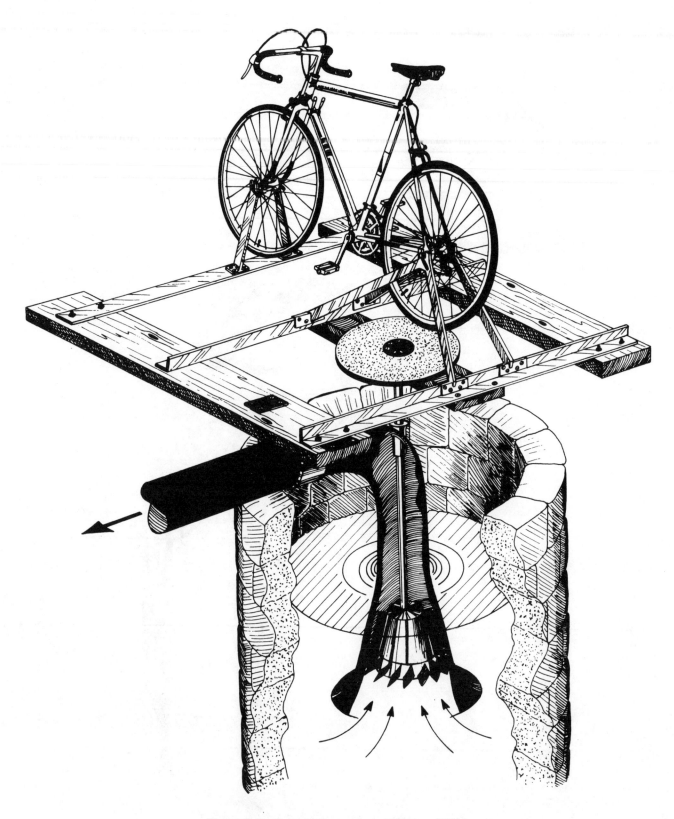

Figure 6-8 Axial water pump (artist's rendition)

All of these are improvements over hand pumps, but none use the leg muscles optimally. Nor do the pumping devices have the highest hydraulic efficiencies available for the duty. For low heads, centrifugal and axial-flow pumps are the most efficient. I once designed and made, for VITA, an axial-flow hydraulic pump which was driven by the rear wheel of my bicycle when the wheel nuts rested in a stand. It was easy to pump between 100 and 200 gallons per minute over a few feet of head, far more than that achievable by other pumps; but the drive was far from successful, and the control system which was needed to cope with different heads and different power inputs simultaneously was not solved. Here is an area which would reward good design.

Tire Pumps

Very few motorists would think of pumping up their own tires nowadays, even in an emergency, and the same is becoming true of bicyclists. Again we find the same story: inefficient, awkward hand pumps requiring body contortions in operation compete unequally with well-designed electric pumps which can, for instance, be plugged into the car cigarette lighter.

We used to have foot pumps at least, which, although they employed only one leg in a nonoptimum manner, could be used in comparative comfort and stored compactly. They were vastly better than hand pumps. I used one connected to a tire to spray-paint my brother's MG during a two-week school vacation. My right leg muscles grew noticeably as a result. An air compressor which was actuated by both legs working in an efficient, continuous motion would undoubtedly make the purchase of a proportion of electrically powered air compressors unnecessary.

Saws

The handsaw is another tool which has scarcely changed in design over the last century, while power saws become ever more convenient and attractive. Handsawing large pieces of wood, such as trees and logs, requires that maximum power be expended by one arm only, often while the body is twisted awkwardly. It is small wonder that we so quickly resort to using portable gasoline- or electric-powered saws, most of which have an output similar to that we could put out ourselves. A pedal-powered table saw, particularly for sawing firewood from logs, is certainly feasible; and a good design would be welcomed by many.

Sewing Machines, Typewriters

Treadle sewing machines were greatly preferred over handcranked models, and typewriters powered by an occasional stroke of a foot have been developed experimentally. The power involved in this class of machine is small, which means that it is easy to produce the power by either muscles or motors. The motors have won out because of convenience and compactness more than for any other reasons. (Obviously an electric typewriter is faster to operate than a manual typewriter, but it would not be faster than a power typewriter where the energy is supplied by the legs.) With the increasing realization that workers in sedentary jobs need some sort of regular mild exercise to keep heart and lungs from being prey to degenerative diseases, it is possible that employers would install reliable and convenient pedal-powered machines. I believe also that many people find, as I do, that the background persistent hum of small motors adds a minor stress to one's existence. Why otherwise should one experience such a wave of relief when a typewriter or an air conditioner is switched off? For this reason alone I would welcome a pedal-powered typewriter.

Figure 6-9 Pedal-operated fan (artist's rendition)

Cooling Fans

Women used to carry Chinese-style fans to cool their faces; men used a newspaper or any other convenient sheet; colonialists in the tropics would have servants to pull and push large, slow-moving, oscillating overhead fans for them. Nowadays we use electric fans for those few places which are not over-cooled by air conditioning.

Now a good circulation is necessary for the body to cope with heat as well as with cold. Obviously, to pedal an exercise ergometer in hot weather is to invite heat stress, but the energy output required to induce a mild cooling current of air over the body is very small. This may be a long shot, but I believe that a form of rocking chair, for instance, in which the legs rocked an oscillating fan might be a welcome addition to a summer terrace or even to a hot living room.

Delivery Vehicles

The days of the huge and heavy delivery-boy's bicycle for meats and groceries have slipped on their unlamented way. There remains an apparent need for some sort of aid for the type of deliveries which at present are made largely on foot. The urban mailman and newspaper boy (and their female counterparts) usually do much of their rounds on foot, carrying heavy shoulder bags. Some mailmen have been equipped with electric golf carts, but the batteries make them heavy and unable to negotiate small hills and steps. Newspaper boys sometimes make their rounds by bicycle, but in urban areas they have to be stationary much of their time, fishing for papers from the ever-present shoulder bag. Delivery people in industrial and commercial plants must still usually push or pull heavy and poorly designed carts.

There seems to be a need for an improved pedaled vehicle. It would be something between a rear-pedaled rickshaw and an ice cream tricycle. It would have multiratio gears, be constructed of ultralight materials, and incorporate some form of rotary (probably) filing or storage in the front carrier, which could be loaded in reverse order of de-

Figure 6-10 Newspaper delivery bicycle (artist's rendition)

livery. I have sometimes thought of such a vehicle being equipped with a spring-loaded howitzer and an aiming telescope, so that the newspaper would have a better-than-even chance of hitting the porch and missing the bushes.

In developing countries, the bicycle or tricycle as a delivery vehicle forms a vital part of a business. The need for improved designs, particularly for reduced weight, better brakes, and efficient, multispeed transmissions, is obvious to any visitor.

Railbike

An intriguing form of bicycle transportation, once very popular in America, is the railbike. Very simply, the railbike is a regular bike fitted with special attachments so that it can be ridden on rails.

In the latter part of the nineteenth century, when ingenious Americans tried to apply principles of the bicycle to all aspects of life, railbike patents flooded the patent office. And some of the designs were quite sound. Not surprisingly, there is evidence that

Figure 6-11 The railbike is a viable form of transportation in many sections of the world.

Figure 6-12

Figure 6-13

Figure 6-14

Figures 6-12 through 6-14
Details of railbike attachments

people traveled hundreds of miles on their railbikes—an interesting and precarious form of transportation.

Today the Bureau of Outdoor Recreation estimates that there are well over 10,000 miles of unused railroad track in America; as more lines fold, unused track will increase. Should these tracks be pulled up, at great expense, or is there a better solution?

Some people believe they have found the solution: the reintroduction of the railbike on unused and little used track. William J. Gillum, president of the American Railbike Association and proponent of widespread railbike use, believes the time has come for the rediscovery of an old idea.

"My interest in rigging up a bicycle that can run on railroad tracks," Gillum wrote in *Harper's Weekly*, "started a couple of years ago. With my friends George and Harris, I was prospecting in the Colorado mountains. We had stopped to rest in a remote and silent area on a railroad track that appeared abandoned. Along the track, from around a bend came a little cart being pushed by a fellow prospector. It had three wheels, two on one rail and one on the other. The tires had little flanges, just like train wheels, which held them on the track. The whole thing folded in the middle, much like a fold-in wheelchair and could be, the Englishman told us, neatly packed in the trunk of his car. He had enough gear piled high on that cart to start mining then and there."

Because of this experience, Gillum and his friends struggled to build a railbike of their own, learning in due course that they were following in the footsteps of hundreds of inventors in the late 1800s and early 1900s—as the old patents reveal.

However, another contemporary inventor, Mark Hansen of Onamia, Minnesota, has enlarged the concept of the railroad bike by building a vehicle which will support two or three people or 700 lbs. He described his railbike in an article in *Alternate Sources of Energy* magazine:

Figure 6-15

Figure 6-16

"The outer passenger wheel is a Schwinn steel front rim. As shown in [Figure 6-15], the passenger is supported by 2-inch channel iron. The horizontal strap around the wheel is ⅛-inch iron strap, and the piece over the top of the wheel is ½-inch conduit. I highly recommend this piece to keep the expansion bridge from twisting.

"As shown in [Figure 6-16], the railroad bike is anchored to the tracks by two boggie wheels. These are 4-inch lawn mower wheels with ball bearings. They are set at 45 degrees and run on both edges of the railway the very point where the front wheel meets the track. The wheels are attached to ⅞-inch square tubing which telescopes into 1-inch tubing. They raise or lower for street or rail use and are held in place with a lock nut. The square tubing is connected to 2-inch channel iron and to a yoke around the front wheel as shown in [Figure 6-15]. This yoke can be locked in place for rail use.

"The driver and passenger units are connected via an expansion bridge as shown in

[Figure 6-15]. This bridge is made from 1-inch conduit. The foremost piece of conduit is welded directly behind the steering column and extends to the front of the passenger wheel harness. This handles the stress that is directed backward when there is a passenger. The rear section of the conduit extends from the rear bicycle axle to the rear side of the passenger wheel harness. This handles side stress. Cross braces between the pieces of conduit add to strength. I suggest using electrical conduit for this purpose because it's very light but strong material and holds a weld without any problems."

As Mark Hansen points out, at this time there are certain safety and legal problems associated with using the railbike. Yet he contends that "it would be interesting to explore the possibilities of making this form of transportation legal since the railroads do rent train wheels for automobile use under some circumstances. Furthermore, there are many miles of abandoned track where a railroad bike could well be used to provide cheap transportation."

Personal Rapid Transit (PRT) Systems

PRT has become a generic name given to the type of people mover which involves a track with constantly moving vehicles or vehicle spaces. One joins such a system by entering a stationary vehicle, bringing it in some way onto an accelerator, and then having it accelerated at the right moment to insert one's vehicle into an unoccupied space on the moving track.

A crude form of such a system is the ski tow or gondola car used in mountain areas. The track speed is slow enough that very rudimentary forms of accelerators are acceptable—often little more sophisticated than a jerk from a chair in one's rump.

The attraction of PRT systems is that with constantly moving vehicles the capacity of a single track is enormous: a fully occupied system could transport well over 10,000 people per hour in a track little wider than a bicycle path, even in a slow-speed (e.g., 15 mph) system. One moving at higher speeds could take more people.

Now even the most enthusiastic devotees of bicycles are honest enough to recognize that the American public is never going to freely choose to undertake journeys of, say, five miles under their own muscle power. The possibility of requiring only a few hundred yards at most of pedaling, and of accomplishing the "line-haul" portion of the trip on a powered track, is, therefore, intriguing. Many people have dreamed dreams and have drawn up plans. A few have made prototypes. A group at Syracuse University has probably gone furthest with several concepts for towing bicycles and tricycles in, generally, covered tracks.

At MIT we have made model systems and crude full-size prototypes of PRT suitable for retirement communities, for new towns, and for recreation areas in particular. We have seen the synchronized accelerator as the principal design problem. Once the vehicles are safely inserted into a stream of vehicles moving at the same speed, they can be driven by a variety of conventional methods, such as cable pulls or synchronous motors driving a type of cog-railroad transmission. We all know from personal experience that the most difficult and stressful part of driving on a busy expressway is the entrance ramp. Will the cars already traveling at high speed on the road give way and make a space for you?

In an automatic system nothing must be left to chance. When the speed of the vehicles on the main track is low (e.g., 15 mph) the need for accuracy is not extreme, and gravity could be sufficiently precise. One would pedal one's tricycle or pedal-car to the access ramp, which would slope down to the main track. This would be in a cutting 8 or 9 feet deep. The vehicle would be held on the slope by a block, which would be removed at the precise time an unoccupied space passed by a sensor a certain way back on the track. One's vehicle would roll down the slope under gravity, joining the track when traveling at just over 15 mph and entering the space to latch onto whatever driving system is employed. Leaving the track would mean choosing an exit ramp to the right, decelerating up the slope, and being pulled the last few yards by some form of tow. Then one would pedal off to the shops or the beach area nearby.

For greater accuracy and versatility, we favor another type of accelerator which we have been developing. Beneath or beside the acceleration lane would run a constantly running metallic screw. The screw threads would start with zero pitch, and from there the thread angle would gradually increase. At the end of the acceleration track a vehicle with a peg or "follower" in the thread would be traveling at precisely the speed of other vehicles on the main track and would be synchronized to fit into an unoccupied space. A similar synchronized variable-pitch screw would be used for the deceleration lane.

Such transportation systems will not become viable in a general way so long as automobiles in towns are so heavily subsidized (they are) and so relatively inexpensive to operate (again, they are). In a national park or retirement community, a system such as those I've briefly described could be given a test which could have great significance for our use of muscle power and for our transportation systems in the future.

Conclusion

Long before the sun melted Icarus's wings, man has longed to fly under his own power, but even Leonardo da Vinci was forced to give up this quest after 16 years. Faced with some irrefutable facts of physics, man has all but ceased trying to imitate the bird. Instead, he has turned his attention to a more favorable part of the anatomy: the legs.

Ever since Henry Kremer, an industrialist, announced a sizable cash prize for the person who piloted the first pedal-driven plane in a figure-eight, under the stipulations laid down by the Royal Aeronautical Society, man has been pedaling for the stars; and, as the accompanying photograph suggests, he has lost little interest in pedal-powered flight.

While there is a fair chance man-powered flight might evolve as a sport and, as enthusiasts hope, an Olympic event, it is likely that pedal power can be employed most productively on planet Earth.

Reports from many developing countries indicate that, with the prohibitive cost of fuel oil, the bicycle is re-emerging as an important factor in transportation and work. The fact is that, under many circumstances, pedal power is as "appropriate" and effective as the low-horsepower internal combustion engine. And, as the bicycle in a sense "liberated" people at the turn of the century, pedal power can liberate millions again. Women, who throughout the world must daily perform difficult tasks by hand, can benefit. Pedal devices which can grind grain, pump water, clear and plow and cultivate the land, generate electricity, lift heavy loads, wash clothes, saw wood, split logs, shred compost, etc., can drastically transform the workplace. Surely pedal power has the potential for humanizing labor.

Pedal power is different things to different people. Maybe the most useful service it can perform in East Africa is to pump water or to grind grain. On the other hand, Americans might be more interested in using pedal power to run a belt sander, to spray paint, or to mix cement.

But perhaps the result is the same. If the numerous applications of pedal power make us less dependent on finite resources, then we satisfy a common objective. And if pedal power extends beyond class and economic lines, we have put geography to rest.

The woman in New South Wales who joyously owns a pedal-operated washing machine could live almost anywhere.

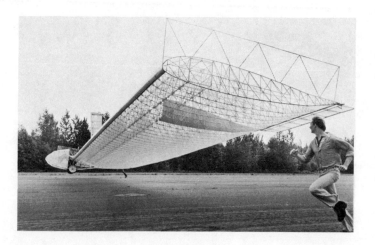

Pedal-powered plane in flight, 1976
(Courtesy of the *Seattle Times*)

Postscript

In some respects pedal power is a fitting symbol of appropriate technology: it can be applied in numerous ways to satisfy a variety of conditions. Yet pedal power, as the subject of serious research, is in the infant stage, where inventions appear to feed the field "by the day."

For that reason, the editor (and contributors) would greatly appreciate ideas and suggestions from readers pertaining to new probes and discoveries in this exciting area.

James C. McCullagh

Appendix

Pedal Power: Research and Development

A bicycle-operated pedal-takeoff unit (plans), ASAE Headquarters, Box 440, St. Joseph, MI 49085. $1.50

Bike-Generator Plans, Homestead Industries, Ananda Village, Nevada City, CA 95959. Drawing and text. $2.00

Bike-Generator Plans, North Shore Ecology Center, 3070 Dato, Highland Park, IL 60035. $1.00

Foot-operated machines, American Village Institute, Route 3, Box 3486, Selah, WA 98942

Oxtrike, Dr. Stuart S. Wilson, Department of Engineering Science, Parks Road, Oxford X 3PJ, England

Reclining Bicycle, Dr. David Gordon Wilson, Department of Mechanical Engineering, MIT, Room 3447, Cambridge, MA 02139

Rodale Energy Cycle, Rodale Resources Division, 33 East Minor Street, Emmaus, PA 18049

Appropriate Technology Groups and Publications

Alternate Sources of Energy (magazine), Route 2, Milaca, MN 56353

Appropriate Technology Center, 82 Allen Hall, Urbana, IL 61801

Brace Research Institute, McDonald College of McGill University, Ste. Anna de Bellevue, P.Q. HOA 1CO, Canada

Intermediate Technology, 556 Santa Cruz Avenue, Menlo Park, CA 94025

Intermediate Technology Publications, Ltd., 9 King Street, London, WC2E 8HN, England. (Publishes *Appropriate Technology*, a quarterly journal, and plans, books, and bibliographies, some of which deal with pedal- and treadle-operated machines in developing countries.)

International Rice Institute, P.O. Box 933, Manila, Philippines

Midwest Appropriate Technology, ACORN, Governors State University, Park Forest South, IL 60466

National Center for Appropriate Technology, Butte, MT 59701

Office of Appropriate Technology, Box 1677, Sacramento, CA 95808

Rain (magazine), 2270 NW Irving, Portland, OR 97210

Tool, Postbus 525, Eindhoven, The Netherlands

TRANET, Transnational Network for Appropriate/Alternative Technologies, 7410 Vernon Square Drive, Alexandria, VA 22306

VITA, Volunteers in Technical Assistance, 3706 Rhode Island Avenue, Mt. Rainier, MD 20822

Volunteers in Asia, Inc., The Clubhouse, Stanford University, Box 4543, Stanford, CA 94305

Bibliography

"A Pedal-Powered Vehicle." *Consumer Reports,* vol. 38, no. 10.

"A Way To Go?" *Popular Science*, vol. 203, August 1973.

"Barrow Inside Giant Wheel Carries Large Loads Easily." *Popular Mechanics*, vol. 71, March 1939.

Beach, Alfred E., ed. *Science Record 1872*. New York: Munn & Co., 1872.

"Bicycle-Canoe Is Driven by Pedal Sidewheel." *Popular Mechanics*, vol. 54, July 1930.

"Bicycle Technology." *Scientific American*, vol. 228, no. 5, May 1973.

Brace Research Institute. *A Handbook of Appropriate Technology*. McDonald College of McGill University, Ste. Anna de Bellevue, P.Q. HOA 1CO, Canada.

Congdon, R. J., ed. *Lectures on Socially Appropriate Technology*. Committee for International Cooperation Activities, Technische Hogeschool, Eindhoven, Netherlands, October 1975.

de Camp, L. Sprague. *The Ancient Engineers*. Cambridge, MA: MIT Press, paperback edition, March 1970.

Hommel, Rudolf P. *China at Work*. John Day Co., copyright 1937; republished Cambridge, MA: MIT Press, September 1969.

"How To Build a Pedal Boat." *Popular Science*, July 1959.

Klemm, Friedrich. *A History of Western Technology*. Cambridge, MA: MIT Press, 1964.

Marks, Vic, ed. *Cloudburst*. Mayne Island, B. C. VON 2JO: Cloudburst Press, Ltd., 1973.

"Motorless Car Pedals Like a Bike." *Popular Mechanics*, vol. 64, November 1935.

"Old Bicycle and Eighty Cents Make a Paddle-Wheel Boat." *Popular Mechanics*, vol. 66, August 1936.

"Pedal Cars: The Gasless Way To Go." *Popular Mechanics*, vol. 141, May 1974.

"Power or Pedal Water Bike." *Popular Mechanics*, vol. 148, August 1965.

Reti, Ladislao, ed. *The Unknown Leonardo*. Maidenhead, UK: McGraw-Hill Book Co. (UK) Ltd.

Ritchie, Andrew. *King of the Road*. Box 4310, Berkeley, CA 94704: Ten-Speed Press (copyright by Andrew Ritchie; Wildwood House Ltd., One Wardour Street, London W1V3HE).

Sharp, Archibald. *Bicycles and Tricycles*. London: Longmans, Green and Co., 1896.

Volunteers in Asia, Inc. *Appropriate Technology Sourcebook.* The Clubhouse, Stanford University, Box 4543, Stanford, CA 94305.

"Ways To Go That You've Never Seen Before." *Popular Mechanics*, vol. 142, October 1974.

Whitt, Frank Rowland, and Wilson, David Gordon. *Bicycling Science*. Cambridge, MA: MIT Press, 1974.

Index